GENETIC MANIPULATION OF RECEPTOR EXPRESSION AND FUNCTION

RECEPTOR BIOCHEMISTRY AND METHODOLOGY

SERIES EDITORS

David R. Sibley
Molecular Neuropharmacology Section
Experimental Therapeutics Branch
NINDS
National Institutes of Health
Bethesda, Maryland

Catherine D. Strader
Department of CNS and
 Cardiovascular Research
Schering-Plough Research
 Institute
Kenilworth, New Jersey

Recent Volumes in Series

Receptor Localization: Laboratory Methods and Procedures
Marjorie A. Ariano, *Volume Editor*

Identification and Expression of G Protein–Coupled Receptors
Kevin R. Lynch, *Volume Editor*

Structure–Function Analysis of G Protein–Coupled Receptors
Jürgen Wess, *Volume Editor*

Regulation of G Protein–Coupled Receptor Function and Expression
Jeffrey L. Benovic, *Volume Editor*

Genetic Manipulation of Receptor Expression and Function
Domenico Accili, *Volume Editor*

Founding Series Editors
J. Craig Venter   Len C. Harrison

GENETIC MANIPULATION OF RECEPTOR EXPRESSION AND FUNCTION

Edited by

DOMENICO ACCILI
Berrie Research Pavillion
Columbia University
New York, New York

WILEY-LISS

A JOHN WILEY & SONS, INC., PUBLICATION

New York • Chichester • Brisbane • Weinheim • Singapore • Toronto

This book is printed on acid-free paper. ⊚

Copyright © 2000 by Wiley-Liss, Inc. All rights reserved.

Published simultaneously in Canada.

While the authors, editors and publisher believe that drug selection and dosage and the specification and usage of equipment and devices, as set forth in this book, are in accord with current recommendation and practice at the time of publication, they accept no legal responsibility for any errors or omission, and make no warranty, express or implied, with respect to material contained herein. In view of ongoing research, equipment modifications, changes in governmental regulation and the constant flow of information relating to drug therapy, drug reactions, and the use of equipment and devices, the reader is urged to review and evaluate the information provided in the package inset or instructions for each drug, piece of equipment, or device for, among other things, any changes in instruction or indication of dosage or usage and for added warnings and precautions.

For ordering and customer service, call 1-800-CALL-WILEY.

Library of Congress Cataloguing-in-Publication Data:

Genetic Manipulation of receptor expression and function / edited by Domenico Accili.
        p. cm. — (Receptor biochemistry and methodology)
    Includes index.
    ISBN 0-471-35057-5 (cloth : alk. paper)
    1. Cell receptors. 2. Genetic engineering. 3. Gene targeting. 4. Cell
receptors—Research—Methodology.   I. Accili, Domenico.   II. Receptor biochemistry and methodology (Unnumbered)

QH603.C43 G47 2000
572.8—dc21   99-042182

Printed in the United States of America.

10 9 8 7 6 5 4 3 2 1

CONTENTS

SERIES PREFACE

The activation of cell surface receptors serves as the initial step in many important physiological pathways, providing a mechanism for circulating hormones or neurotransmitters to stimulate intracellular signaling pathways. Over the past 10–15 years, we have witnessed a new era in receptor research, arising from the application of molecular biology to the field of receptor pharmacology. Receptors can be classified into families on the basis of similar structural and functional characteristics, with significant sequence homology shared among members of a given receptor family. By recoginizing parallels within a receptor family, our understanding of receptor-mediated signaling pathways is moving forward with increasing speed. The application of molecular biological tools to receptor pharmacology now allows us to consider the receptor–ligand interaction from the perspective of the receptor as a complement to the classic approach of probing the binding pocket from the perspective of the ligand.

Against this background, the newly launched Receptor Biochemistry and Methodology series will focus on advances in molecular pharmacology and biochemistry in the receptor field and their application to the elucidation of the mechanism of receptor-mediated cellular processes. The previous version of this series, published in the mid-1980s, focused on the methods used to study membrane-bound receptors at that time. Given the rapid advances in the field over the past decade, the new series will focus broadly on molecular and structural approaches to receptor biology. In this series, we interpret the term *receptor* broadly, covering a large array of signaling molecules including membrane-bound receptors, transporters, and ion channels, as well as intracellular steroid receptors. Each volume will focus on one aspect of receptor biochemistry and will contain chapters covering the basic biochemical and pharmacological properties of the various receptors, as well as short reviews covering the theoretical background and strategies underlying the methodology. We hope that the series will provide a valuable overview of the status of the receptor field in the late 1990s, while also providing information that is of practical utility for scientists working at the laboratory bench. Ultimately, it is our hope that this series, by pulling together molecular and biochemical information from a diverse array of receptor fields, will facilitate the integration of structural and functional insights across receptor families and lead to a broader understanding of these physiologically and clinically important proteins.

David R. Sibley
Catherine D. Strader

PREFACE

Our ability to manipulate the genome of living organisms to understand gene function in vivo has changed dramatically over the course of the past decade. Thanks to numerous advances in the field of transgenic technology, investigators are now in a position to effect virtually any change in the mouse genome. Receptor research has not been immune to these changes. Using transgenesis and homologous recombination in embryonic stem cells, scientists have been able to address long-standing questions in receptor biology. Because of the complex nature of the technology involved, many investigators find it increasingly difficult to keep up with the brisk pace of discovery in transgenic research. This volume is designed to offer a road map to transgenic and knockout methodology to a broad range of investigators, scholars, students, and teaching faculty in biomedical institutions.

Because of its technical complexity, the generation of transgenic and knockout mice hardly lends itself to a cook-book format. Therefore, rather than providing detailed protocols to generate one's own transgenic or knockout mouse, this volume is designed to give the reader a better sense of the questions that can be addressed when using these methods. Equal emphasis is therefore placed on principles of transgenic/knockout methods and applications. The volume is designed to provide the necessary conceptual backdrop for investigators who are new to the field so that they will feel comfortable understanding the difference between a transgenic and a knockout mouse, between embryonic stem cells and fertilized zygotes, between a DNA construct to make a transgenic mouse, and a targeting vector for a knockout (Homanics, Merlino). At the same time, we have attempted to review the considerable technical progress that has been made in generating inducible and tissue-specific mouse models (Comoglio, Drago, Kahn and Kappen). Finally, we have emphasized rapidly changing aspects of transgenic technology, such as the generation of transgenic mice by using large DNA fragments (Camper), the role of modifier genes in determining a phenotype (Magnuson), the importance of behavioral testing in analyzing disease phenotypes (Crawley), the establishment of cell lines from mice (Chou), and the role of antisense techniques in studies of gene function (Malbon and Creese).

Working on the planning and realization of this book has been an exceedingly pleasant experience, and I am especially proud of the end product. The Authors deserve all the credit for bringing to life a tool that will be most useful for scholars and practitioners alike. The Series Editors, Dave Sibley and Catherine Strader, have been of great help in charting my course, and I wish to extend to them a heartfelt thank you.

DOMENICO ACCILI

CONTRIBUTORS

SCOTT BULTMAN, Department of Genetics, Case Western Reserve University, Cleveland, OH

SALLY A. CAMPER, Department of Human Genetics, University of Michigan Medical School, Ann Arbor, MI

GIULIA CELLI, Laboratory of Molecular Biology, National Cancer Institute, Bethesda, MD

JANICE YANG CHOU, Heritable Disorders Branch, National Institute of Child Health and Human Development, National Institutes of Health, Bethesda, MD

PAOLO M. COMOGLIO, Institute for Cancer Research (IRCC), University of Torino Medical School, Candiolo, Torino, Italy

JACQUELINE N. CRAWLEY, Section on Behavioral Neuropharmacology, Experimental Therapeutics Branch, Intramural Research Program, National Institutes of Health, Bethesda, MD

IAN CREESE, Center for Molecular and Behavioral Neuroscience, Rutgers, The State University of New Jersey, Newark, NJ

JOHN DRAGO, Department of Anatomy, Neurosciences Unit, Monash University, Clayton, Australia

SILVIA GIORDANO, Institute for Cancer Research (IRCC), University of Torino Medical School, Candiolo, Torino, Italy

DAVID GRANDY, Department of Cell and Developmental Biology, Oregon Health Sciences University, Portland, OR

PATRICIA GREEN, Department of Genetics, Case Western Reserve University, Cleveland, OH

GREGG E. HOMANICS, Departments of Anesthesiology/CCm and Pharmacology, University of Pittsburgh, PA

C. RONALD KAHN, Research Division, Joslin Diabetes Center and Department of Medicine, Harvard Medical School, Boston, MA

CLAUDIA KAPPEN, Department of Biochemistry and Molecular Biology and Molecular Neuroscience Program, Samuel C. Johnson Medical Research Center, Mayo Clinic Scottsdale, Scottsdale, AZ

SIMRANJIT KAUR, Center for Molecular and Behavioral Neuroscience, Rutgers, The State University of New Jersey, Newark, NJ

ROHIT N. KULKARNI, Research Division, Joslin Diabetes Center and Department of Medicine, Harvard Medical School, Boston, MA

WILLIAM J. LAROCHELLE, Laboratory of Molecular Biology, National Cancer Institute, Bethesda, MD

TERRY R. MAGNUSON, Department of Genetics, Case Western Reserve University, Cleveland, OH

CRAIG C. MALBON, Department of Molecular Pharmacology-HSC, Diabetes & Metabolic Diseases Research Center, School of Medicine, SUNY/Stony Brook, Stony Brook, New York

GLENN MERLINO, Laboratory of Molecular Biology, National Cancer Institute, Bethesda, MD

M. DODSON MICHAEL, Research Division, Joslin Diabetes Center and Department of Medicine, Harvard Medical School, Boston, MA

CHRISTOPHER M. MOXHAM, Department of Molecular Pharmacology-HSC, Diabetes & Metabolic Diseases Research Center, School of Medicine, SUNY/Stony Brook, Stony Brook, New York

THOMAS L. SAUNDERS, Departments of Human Genetics and Medical Administration, University of Michigan Medical School, Ann Arbor, MI

HSIEN-YU WANG, Department of Physiology & Biophysics, Diabetes & Metabolic Diseases Research Center, School of Medicine, SUNY/Stony Brook, Stony Brook, New York

JOHN Y. F. WONG, Department of Anatomy, Neurosciences Unit, Monash University, Clayton, Australia

CHAPTER I

TRANSGENIC RESCUE OF MUTANT PHENOTYPES USING LARGE DNA FRAGMENTS

SALLY A. CAMPER, PH.D. and THOMAS L. SAUNDERS, PH.D.

I. Introduction

In a few years the human and mouse genomes will be completely sequenced, providing researchers around the world with a deluge of information. The DNA sequence will facilitate the identification of genes, but understanding gene function will still be a challenge. Mouse mutations will undoubtedly be important players in the coming era of "functional genomics." Historically, positional cloning has been used to identify the genes responsible for mouse mutations with interesting visible phenotypes. In many cases the discovery of the mouse gene led directly to the identification of the cause of human genetic disease, validating the importance of mouse models (Bedell et al., 1997; Meisler, 1996). Mutagenesis in the mouse will generate many more mutants for analysis of gene function. This process will involve directed mutagenesis by homologous recombination in embryonic stem (ES) cells to a certain extent, but a great deal of information will be gained from saturation mutagenesis of genomic regions by using ES cell technology and chemical mutagenesis (Schimenti and Bucan, 1998).

Mice offer several advantages over the analysis of human pedigrees for identification of mutant genes. Genetically distant inbred strains of mice can be used to generate a large "family" of informative recombination events. For example, 500 F_1 intercross progeny can be readily generated and analyzed, permitting rapid localization of a mutant gene to <1 cM, a smaller genetic interval than is feasible with most human pedigrees. However, the size of the physical interval that must be searched for the mutant gene is often daunting, as 1 cM can represent >2000 kb. Genetic complementation with either cell cultures or transgenic mice offers an alternative to narrowing the physical interval that must be searched for genes. To be successful, the sequences required for transcription and coding of the gene must be contained in the genomic clone used

Genetic Manipulation of Receptor Expression and Function,
Edited by Domenico Accili.
ISBN 0-471-35057-5 Copyright © 2000 Wiley-Liss, Inc.

for complementation. If cloning vectors that accept large DNA inserts are used, it is more likely that all of the regulatory and structural information required for functional gene expression will be present in single genomic clones. The combination of positional cloning with genetic complementation can significantly shorten the time needed for gene identification.

Advances in engineering the mouse genome have ensured that the mouse will play an important role in developing new models for human genetic disease. These models are being used to test the efficacy of gene therapy and conventional pharmaceutical therapies for heart disease, cystic fibrosis, muscular dystrophy, sickle-cell anemia, and neurologic disease (Manson et al., 1997; Rozmahel et al., 1997; Hauser et al., 1997; Paszty et al., 1997; Kling et al., 1997; Veniant et al., 1999). Mouse models for human disease can often be generated by homologous recombination in ES cells, but some disease models require introduction of human genes into the mouse. Because the DNA sequences necessary for physiological levels of expression can lie a considerable distance from a gene, it is essential to be able to introduce large genomic clones into the mouse germline. Recently, examples of successful production of transgenic mice with large genomic clones have become more common.

2. Overview of Mouse Positional Cloning and Mouse Genomic Libraries

The steps in positional cloning are genetic mapping, physical mapping (cloning all of the DNA between the nearest proximal and distal genetic markers), gene identification (cDNA selection, exon trapping, and/or DNA sequence analysis with computer-aided gene identification), mutation detection, and correlation of the mutant gene with the observed phenotype (see Boehm, 1998 for review). Construction of the genetic map has been facilitated by the development of simple sequence repeat markers that are highly polymorphic among inbred strains of mice and readily typed by PCR analysis (Dietrich et al., 1995). Recently, physical mapping has become easier, as large genomic clones (YACs or yeast artificial chromosome clones) have been isolated and characterized by assaying their content of simple sequence repeat markers (Haldi et al., 1996). These genomic clones can be purchased and used as a starting point for the physical map. The time spent for gene identification and mutation detection can be reduced if the physical interval to be searched is narrowed by genetic complementation.

The average gene is ~30 kb, although gene size can range from <2 kb to >1 Mb. Very large genes will probably be extremely difficult to identify by complementation, but the majority should be tractable with existing mouse genomic libraries. Several types of vectors are available and have been used for complementation analysis (Tables 1.1 and 1.2). The largest clones are YACs, which contain 100 to 1000 kb of DNA or more. The utility of YACs is limited by the high rates of chimerism (clones with multiple inserts) and instability, the difficulty of handling such large DNA without shearing, and the problem of purifying YAC DNA from yeast DNA. If YACs are used for complementation it is important to verify the integrity of the clones before and after injection. Bacteriophage P1 clones are much easier to work with than YACs, but their small size (70 kb) and low copy number are disadvantages. The likelihood of having sufficient sequences on an individual P1 clone for complementation is lower than with larger clones, and purification of DNA can be problematic. Bacterial artificial chromosome clones (BACs) generally have larger inserts (100–150 kb) than P1 clones, are easier to purify, and are reportedly quite stable. Also noteworthy is the recent development of methods to modify BACs by homologous recombination in bacteria, enhancing the utility of BAC transgene technology (Yang et al., 1997; Zhang et al., 1998). PAC (P1-derived artificial chromosomes) vectors are another alternative. They have an average insert size of 130–150 kb (Ioannou et al., 1994). We have had the most experience with BAC clones and recommend them for complementation analysis.

3. Overview of Transgenic Mouse Production

Transgenic technology became a reality in the early 1980s. The ability to introduce artificially assembled genes into the mouse genome advanced knowledge in many different fields. Comprehensive review of this topic is beyond the scope of this chapter, but a few citations are given as examples. The basic parameters for transgenic mouse production were described by Brinster et al. (1985). Standardized methods were described by Hogan et al. (1994). The original methods used to produce transgenic mice have not changed substantially since they were introduced (Palmiter et al., 1982). The production of transgenic mice involves the isolation of fertilized eggs, the microinjection of soluble DNA into pronuclei, and the transfer of the injected eggs to a surrogate mother. The unique aspects of a given experiment are genetic cloning of the transgene and analysis of transgene expression in animals (Fig. 1.1).

Transgenes have two critical components, the transcription unit—including exons, introns, and polyadenylation sequences—and the promoter and enhancer sequences that direct expression of the gene. Generally, transient expression of genes by transfection of cell cultures can be achieved with very short promoter sequences of 200 bp or less. However, often more sequence is necessary to obtain expression in intact animals. The generation of transgenic mice with large DNA clones such as P1s, YACs, and BACs has been an important step in the identification of DNA sequences important for gene expression. In addition, such large DNA clones have begun to be used in genetic complementation experiments (Tables 1.1 and 1.2, see below).

Table 1.1. Identification of Genes by Genetic Complementation: in vitro Transgenesis

Mutation	Critical Region Before Complementation	Critical Region After Complementation	Reference
H-3a	2 cm	single YAC	Zuberi et al., 1998
Niemann-Pick Type C	1.2 Mb	300 kb	Gu et al., 1997
Fv1	1.2 Mb	190 kb	Best et al., 1996
beige	2.4 Mb	650 kb	Barbosa et al., 1996
beige	1.0 Mb	500 kb	Perou et al., 1996

Table 1.2. Identification of Genes by Genetic Complementation in vivo Transgenesis

Mutation	Critical Region Before Complementaion	Critical Region After Complementation	Cloning Vector	Vector Included	DNA Form	Transgenic % Born	Efficiency % Transgenic	Copy Number	Reference
downless	0.6 mb	200 kb	YAC	Yes	Native	7%	33%	5 to 10	Madjumder, et al., 1998
shaker-2	1.0 mb	140 kb	BAC	No	Linearized	16%	12%	—*	Probst, et al., 1998
Clock	0.4 mb	140 kb	BAC	Yes	Circular	—*	9%	1 to 5	Antoch, et al. 1997
				No	Linearized	—*	6%	2 to 12	
vibrator	0.5 mb	76 kb	P1	Yes	Circular	—*	—*	—*	Hamilton, et al., 1997
minibrain	2.0 mb	180 kb	YAC	Yes	Native	—*	—*	1 to 3	Smith, et al., 1997

*Not reported

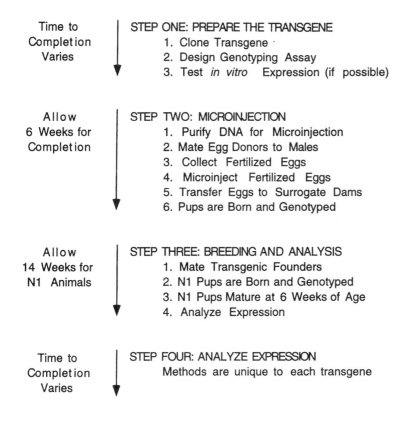

Time to
Completion
Varies

STEP ONE: PREPARE THE TRANSGENE
  1. Clone Transgene
  2. Design Genotyping Assay
  3. Test *in vitro* Expression (if possible)

Allow
6 Weeks for
Completion

STEP TWO: MICROINJECTION
  1. Purify DNA for Microinjection
  2. Mate Egg Donors to Males
  3. Collect Fertilized Eggs
  4. Microinject Fertilized Eggs
  5. Transfer Eggs to Surrogate Dams
  6. Pups are Born and Genotyped

Allow
14 Weeks for
N1 Animals

STEP THREE: BREEDING AND ANALYSIS
  1. Mate Transgenic Founders
  2. N1 Pups are Born and Genotyped
  3. N1 Pups Mature at 6 Weeks of Age
  4. Analyze Expression

Time to
Completion
Varies

STEP FOUR: ANALYZE EXPRESSION
  Methods are unique to each transgene

Total Time to Complete Project: 20 Weeks After Transgene Preparation

Figure 1.1. Timeline for Production and Analysis of Transgenic Mice.

If an appropriate cell culture system is available, the transgene can be tested for expression before animals are generated. Although expression in cell culture is not a guarantee of expression in intact animals, cell culture experiments can be valuable for identifying problems in transgene construction or in detection of transgene expression. After the transgene has been cloned and purified, it is microinjected into fertilized mouse eggs (Fig. 1.2). Most often F_2 hybrid eggs are used as hybrid vigor enhances the efficiency of transgenic mouse production. Eggs that survive the microinjection process are transferred to pseudopregnant female mice that develop the eggs to term and raise the pups to weaning age (Fig. 1.1). Of the pups that are born, a certain proportion will be transgenic, ranging from 0% to 45% of the offspring. If no transgenic progeny are obtained, the transgene may be lethal. In most cases, however, this outcome results from poor preparation or quantitation of the DNA or from insensitive transgene detection methods. Ideally, if one begins with 100 eggs, ~80% will be fertilized; 80% will survive microinjection; and 80% of the unlysed eggs develop to two-cell embryos, and ~20% of these will be born as pups, and ~20% of the pups (two pups) will be transgenic. Reduced efficiency of any one of these factors will have serious negative consequences on transgenic production. For example, use of mutant or inbred mouse strains as egg donors reduces the efficiency dramatically (Brinster et al., 1985). In some cases, <10% of the eggs produced are fertilized (Saunders and Camper, unpublished observations).

After microinjection, transgene DNA integrates randomly into the mouse genome. If this integration happens before extensive cell division, the transgene will be present in every cell of the founder animal. When the founder animal is bred to a nontransgenic animal, the transgene will be transmitted in a Mendelian fashion to 50% of the progeny (Fig. 1.1). Occasionally the transgene integrates after a substantial number of cell divisions, resulting in a mosaic animal that will transmit the transgene poorly, if at all. Typically, transgene integration occurs as a multicopy head-to-tail concatamer of DNA molecules found on one or, occasionally, two chromosomes (Bishop and Smith 1989). The presence of the transgene can be detected by PCR or Southern blot analysis. PCR detection is fast, but Southern blot analysis is valuable for verification of transgene integrity, determination of copy number, and detection of multiple chromosomal integration sites (Camper 1985). In addition, false positives are rare by Southern analysis.

4. Regulatory Sequences Often are a Significant Distance from the Transcription Start Site

In transgene design it is important to consider whether sufficient genetic information is present in the cloned DNA fragment to obtain the desired gene expression, including tissue specificity, developmental regulation, and adequate expression level. For genetic complementation this information is critically important as levels of lower-than-normal endogenous gene expression may not be sufficient for phenotypic correction. Locus control regions (LCR) are genetic regulatory sequences that control accessibility of the DNA to transcription factors. These sequences can be located tens of kilobases away from coding exons, and they can confer copy-number-dependent and site-of-integration-independent expression on transgenes. (Peterson, 1997; Nielsen et al., 1998). Most transgenes are not large enough to contain locus control regions, and heterologous LCRs incorporated by genetic engineering are not always reliable (Keegan et al., 1994). Therefore the level of expression of most transgenes is influenced more by the site of transgene integration than by the copy number, necessitating analysis of multiple lines with each construct.

Large genomic clones are ideal for complementation experiments because regulatory information may be present in intronic sequence or at a considerable distance from a gene. The importance of intronic sequences for expression is illustrated by attempts to correct severe intestinal dysfunction caused by a defective cystic fibrosis transmembrane regulator (CFTR) gene. A YAC transgene containing the human CFTR gene produced sufficient protein to correct the intestinal phenotype of CFTR null mice (Manson et al., 1997). In contrast, a human CFTR cDNA sequence placed under the control of the endogenous mouse CFTR 5' sequences using a "knock-in" strategy of homologous recombination in ES cells (Rozmahel et al., 1997) did not rescue the phenotype. It is interesting in this case that rescue was achieved by the transgene, even though the normal expression patterns of CFTR were not precisely reproduced. This scenario leaves open the possibility that additional elements are required for expression that were not contained within the YAC. The failure of the "knock-in" to correct the phenotype may be due to the deletion of the intronic sequences.

Efforts to establish a model of atherosclerotic disease through expression of human lipoprotein a (apoB100) provides another example of the value of including as much DNA sequence as possible for optimal gene expression (see review by McCormick and Nielsen, 1998). Three groups designed various transgene constructs with human apoB100 cDNA and collectively generated about 40 transgenic founder mice (Chiesa et al., 1993; Scott et al., 1989; Xiong et al., 1991, 1992). None of the constructs were very successful. Only one transgenic line expressed the apoB100 protein. Subsequently, a P1 clone with 79.5 kb of human sequence, including the apoB100 gene, was used to generate transgenic mice (Linton et al., 1993). The P1 contained all of the intronic sequences, 19-kb of exon 1 and 17.5-kb of the polyadenylation site. The majority of the transgenic founders expressed physiological levels of apoB100 (27/30 or 82%), indicating that the large genomic clone was much more effective than smaller cDNA constructs for the expression of apoB100 (1/40 or 2.5%). Despite the high penetrance of expression of the P1 transgene, the tissue-specific pattern of expression was not normal. Liver expression was high, but no expression was detected in the intestine (Linton et al., 1993, Callow et al., 1994). A larger clone (145-kb BAC) expressed apoB100 in both liver and intestine of transgenic mice. The elements required for intestinal expression were located 54 to 62 kb upstream of the gene, whereas the elements required for liver expression were localized to the region 5-kb upstream of the gene (Nielsen et al., 1998).

Most of the complementation studies that have successfully identified genes by introducing genomic DNA into culture systems or animals have relied on large genomic fragments cloned in YAC, BAC, or P1 vectors. The use of

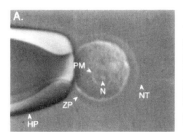 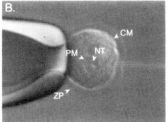 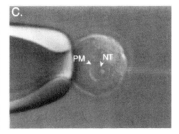

Figure 1.2. Pronuclear Microinjection of Fertilized Mouse Eggs. A. The egg is held in place with a holding pipet (HP). The zona pellucida (ZP) prevents the cytoplasm from being aspirated into the holding pipet. The pronuclear membrane (PM) is brought into focus, and the tip of the injection needle (NT) is raised or lowered until it is in the same plane of focus as the pronuclear membrane. A single nucleolus (N) is visible inside the pronucleus. B. The tip of the injection needle is guided into the pronucleus. The zona pellucida (ZP), cytoplasmic membrane (CM), and pronuclear membrane (PM) are pierced. C. DNA in solution is injected into the pronucleus. Successful injection occurs when the pronucleus swells visibly and has an increased diameter.

smaller genomic sequences, such as those found in lambda or cosmid libraries, would decrease the probability of genetic complementation. The apoB100 example illustrates a situation in which the structural gene is 45 kb, but inclusion of the upstream regulatory elements requires more than 100 kb, a size range that can be accommodated by YAC or BAC clones.

5. Complementation in Cell Culture

Complementation of genetic defects in cultured cells is an effective means of gene identification. Cells offer the advantage of low cost and rapid analysis compared with the use of animals as transgenesis targets. The ease with which cells can be maintained in tissue culture and their amenability to yeast spheroplast fusion makes their use in complementation attractive. There is a growing list of success stories in which a disease gene was identified by using complementation in cell culture. Genes responsible for retrovirus restriction (Best et al., 1996), the Chediak-Higashi syndrome (Perou et al., 1996), the Niemann-Pick C disease (NPC1) (Carstea et al., 1997), and a histocompatibility antigen (Zuberi et al., 1998) have all been identified by this approach. A brief discussion of each of these examples illustrates the strengths and limitations of complementation in cell culture (Table 1.1).

Identification of the Fv1 retrovirus restriction gene relied on the ability of genomic clones to render L cells resistant to infection by B-tropic murine leukemia viruses. This example illustrates the strength of complementation in the identification of a small gene. In this case the identification of transcribed sequences was completely obviated (Best et al., 1996). A YAC that endowed L cells with virus resistance was used to construct a cosmid library. A complementing cosmid was identified and subcloned. The subclones were tested for complementation. This process narrowed down the region to a 6.5-kb DNA fragment, which was sequenced. A 1.4-kb open reading frame was identified and shown to complement when introduced into the cells with an expression vector. Thus, the rapid complementation assay permitted a number of successively smaller clones to be assessed, speeding the time to identify the gene.

The mouse *beige* mutation is a model of Chediak-Higashi syndrome in humans. Mutants have abnormal vesicle morphology, such as unusually enlarged lysosomes and melanosomes. Lysosome dysmorphology is apparent in cultured fibroblasts from the mutant, whereas the abnormal melanosomes result in a hypopigmentation phenotype in the mouse. Although complementation could have been carried out in animals, spheroplast fusion of YACs in cell culture was the method of choice by two groups (Barbosa et al., 1996, Perou et al., 1996). Although the YAC integrity was not perfect, complementation was still achieved, culminating in the description of the human homologue of the *beige* gene, LYST, which was found to be mutated in human patients suffering from Chediak-Higashi syndrome (Barbosa et al., 1996).

The Niemann-Pick type C (NP-C) disease gene, like the *beige* mutation, causes a metabolic disorder that can be observed with cultured cells. A Chinese hamster ovary cell line (CT60) with cholesterol accumulation characteristic of the NP-C phenotype was used for complementation studies. Four human

YAC clones covering 1.5 Mb spanned the region identified by genetic mapping. Only one of the YACs complemented the mutation. By excluding overlapping regions with other YACs, the critical region was narrowed to 300 kb. Exons were mapped in four human BAC clones that spanned the critical interval. A cDNA expression construct was used to confirm the identification of the NP-C gene by phenotypic correction in cultured cells. This research led to the identification of the gene responsible for NP-C disease in humans (Carstea et al., 1997).

A positional cloning approach was used to identify the H3a minor histo-compatibility antigen gene (Zuberi et al., 1998). Genetic mapping identified a 1–2 cM region that contained H3a, and the critical region was cloned in YACs. The physical size of this region could not be determined readily because the majority of the YACs were chimeric (31/36). A cytotoxic T-cell line (CTL) that lyses cells expressing H3a was used to determine which of the YACs contained the H3a gene. D^b L-cells are not normally lysed by the CTL, but susceptibility to lysis was conferred by spheroplast fusion with yeast carrying the YAC with the H3a gene. Five P1 clones that mapped to the YAC were identified and tested for complementation (Zuberi et al., 1998). The sequence of the complementing P1 clone was determined and used to identify candidate genes. A gene that ex-hibited the expected tissue specificity was able to confer H3a expression in transfected cells. Like the Fv1 gene identification, successive reduction in clone size led to rapid identification of the critical gene.

6. Genetic Complementation in Whole Animals

It is clear that complementation studies are limited by the availability of an ap-propriate cell line. For example, behavioral phenotypes associated with circa-dian rhythm, learning, hearing deficiency, and neurological defects require complementation in intact animals (Antoch et al., 1997; Hamilton et al., 1997; Madjumder et al., 1998; Probst et al., 1998). The following examples of com-plementation illustrate the use of large DNA clones to identify mutations in whole animals (Table 1.2).

The *clock* mutation disrupts normal circadian rhythm in the mouse. Identi-fication of this gene was one of the first examples of complementation using transgenic mice. A genetic map was constructed by using 2400 meioses, nar-rowing the critical region to 400 kb (King et al., 1997). A physical map of 32 YACs was assembled. Because half of the YACs were chimeric, the YAC clones were used to identify 12 BAC clones from the region. Fine mapping with mark-ers derived from the BACs reduced the critical region to 250 kb. Transgenic mice were generated from overlapping DNA fragments contained in two BACs. Fertilized eggs that were wild type or heterozygous for the dominant *clock* mu-tation were microinjected. One of the intact 140 kb BACs rescued *clock* mutant mice, restoring the normal circadian rhythm. Deductive reasoning based on knowledge of the BAC fragments that failed to complement and the overlap of the fragments tested permitted the critical region to be narrowed to 40 kb. DNA sequence analysis and transcript mapping identified a gene partially encoded in this region. It was expressed at lower levels in mutant mice. A point mutation in a splice donor site caused the deletion of 51 amino acids. Identification of the *clock* gene was greatly facilitated by the complementation approach.

The *vibrator* mutation causes a neurodegenerative disease. The gene was identified by complementation by using transgenic mice (Hamilton et al., 1997). A genetic map was constructed by using recombination information from 2600 meioses. A single YAC that covered the nonrecombinant interval was used to develop sequence tagged sites (STS) for finer mapping and isolation of 29 P1 clones. The physical distance was estimated to be ~500 kb, but many genes were identified in this interval by cDNA selection. Five strong candidate genes with appropriate expression patterns were identified, but no mutations were identified in *vibrator* mice. Complementation was used to restrict the number of genes under consideration. The eggs used for microinjection were obtained by mating wild-type female mice with males heterozygous for the recessive mutation. A 76-kb P1 clone was found to complement the *vibrator* mutation. Transgenic mice lacked the characteristic tremors of *vibrator* mice and had an increased lifespan. DNA sequencing revealed that the P1 clone contained four genes. Whereas each of the genes was expressed in *vibrator* mice, *Pitpn*—a putative phosphoinositol phosphatase gene—was expressed at fivefold lower levels, implicating it as the critical gene. No coding region mutations were found, but an intracisternal A particle insertion into an intron of the *Pitpn* was found to be the cause of reduced expression. Thus, complementation effectively focused the search for the gene with much higher resolution than was possible with a large backcross.

The *shaker-2* mutation causes recessive congenital deafness and vestibular defects that result in head-tossing and circling behavior. Comparative genetic mapping suggested that *shaker-2* was the mouse homolog of the human deafness locus *DFNB3* (Liang et al., 1998). Identification of the mouse gene was facilitated by gene complementation in transgenic mice, which contributed to the identification of the human gene and several deafness-causing mutations within it (Probst et al., 1998; Wang et al., 1998). Genetic mapping and physical mapping narrowed the *shaker-2* critical region to ~1 Mb of YAC and BAC clones, whereas intensive human genetic mapping resolved the interval to only ~3 cM (Wakabayashi et al., 1997; Probst et al., 1999; Liang et al., 1998). BAC clones were tested for complementation by making transgenic mice directly in mutant *shaker-2* fertilized eggs. The eggs were obtained by mating homozygous *shaker-2* females with homozygous *shaker-2* male mice. Phenotypic correction by microinjection of a 140-kb BAC transgene was observed as normal locomotion and normal startle response to sound. The BAC was sequenced and gene identification with computer algorithms (GENSCAN, GRAIL, and BLAST) revealed a number of putative exons encoding a novel myosin protein (*Myo15*). Comparison of the mouse and human genomic sequences was valuable for identification of potential coding regions. Point mutations were found in *shaker-2* mice and three unrelated *DFNB3* pedigrees, confirming that the correct gene had been identified.

Transgene correction was used to locate the *downless* mutation (Madjumder et al., 1998). *Downless* mice have a defect in hair-follicle development that causes resulting baldness and hair abnormalities (Shawlot et al., 1989). An allele generated by transgene insertional mutagenesis was used to identify genomic sequences near or within the mutant gene, and these sequences were used to clone 1 Mb in YACs. The YACs were microinjected into eggs that were heterozygous or homozygous for the *downless* mutation. A 200-kb YAC was found to correct the phenotype in the mice with the transgene-induced deletion. Although the gene involved in this mutation has yet to be identified, the genomic region for further study has been reduced from 1000 kb to 200 kb by complementation, expediting the path to gene discovery.

Generation of an animal model for Down's syndrome, a common human trisomy, with YAC transgenes illustrates an alternative use for transgenes in gene identification (Smith et al., 1997). The experiments above all relied on a transgene to correct the phenotype of existing dominant or recessive mouse mutations. In this case, YAC transgenic mice were used to generate a mutant phenotype similar to that exhibited by humans with Down's syndrome, providing the first step towards identifying the critical dosage-sensitive genes that cause the phenotype. A 570-kb YAC from a 2-Mb physical map was found to confer learning defects to transgenic mice. Transgenic mice generated with fragments of this YAC narrowed the interval to 180 kb. The human homolog of the *Drosophila minibrain* gene is contained in the fragmented YAC. Appropriate expression of this 100 kb YAC in transgenic mice generated learning deficits, which suggests that it contributes to Down's syndrome. By this method, other dosage-sensitive genes could be identified.

7. Technical Considerations for Efficient Production of Transgenic Mice with Large Genomic Clones

The same process is used to produce transgenic animals with large and small DNA fragments. DNA is microinjected into pronuclei, surviving eggs are transferred to pseudopregnant foster mothers, and DNA from pups is analyzed for the presence of the transgene (Fig. 1.1, Hogan et al., 1994). The overall efficiency of generating transgenic mice is lower with large DNA fragments because of the lower birth rate. The proportion of mice born that are transgenic is comparable (Table 1.3). The technical aspects of transgenesis with large and small DNA molecules are compared in Table 1.4. The most significant difference between producing transgenic mice with large and small DNA fragments is that large DNA is more susceptible to fragmentation during preparation, storage, and microinjection. For example, based on our experience with hundreds of transgenes 20 kb or less in size, Southern blot usually confirms the presence of intact transgenes. In contrast, transgenic mice produced by microinjection of large DNA fragments often contain incomplete fragments whether the DNA is obtained from a YAC (Schedl et al., 1993; Smith et al., 1995), BAC (Saunders and Camper, unpublished observations), or P1 clone (Smith et al., 1995). The use of an injection buffer supplemented with salt and polyamines improves the yield of intact YAC, BAC, and P1 transgenes (Schedl et al., 1993; Antoch et al., 1997; McCormick et al., 1994; Saunders and Camper, unpublished observations). This modified buffer (10-mM Tris, pH 7.5, 0.1-mM EDTA, 100-mM NaCl, 30-μm spermine, 70-μm spermidine) results in compact, globular DNA structures estimated to be 1 μm in diameter and protects against breakage

Table 1.3. Transgenic Efficiency of Large Insert Cloning Vectors

Mutation or Gene	Cloning Vector	Transgene Size	Vector Included[a]	DNA Form	Transgenic % Born[b]	Efficiency % Transgenic[c]	Copy Number[d]	Expressing Founders[e]	Reference
shaker-2	BAC	140 kb	Yes	Linear	16%	12%	—	1/1	Probst et al., 1998
Clock	BAC	140 kb	Yes	Circular	—	9%	1 to 5	2/2	Antoch et al., 1997
	BAC	100 kb	No	Linear	—	6%	2 to 12	—	
RU49	BAC	131 kb	Yes	Linear	—	20%	1 to 3	1/1	Yang et al., 1997
apoB	BAC	68 to 124 kb	No	Linear	—	15%	1 to 4	22/22	Nielsen et al., 1998
apoB	P1	80 kb	No	Linear	—	—	1 to 42	16/20	Linton et al., 1993
apoB	P1	80 kb	Yes	Both	—	28%	1 to 15	11/11	Callow et al., 1994
vibrator	P1	76 kb	Yes	Circular	—	—	—	2/2	Hamilton et al., 1997
human IgH	P1	180 kb[f]	No	Linear	—	12%	1	1/2	Wagner et al., 1996
minibrain	YAC	180 kb	Yes	Circular	—	—	1 to 3	—	Smith et al., 1997
downless	YAC	200 kb	Yes	Circular	7%	33%	5 to 10	3/3	Majumder et al., 1998
β globin	YAC	248 kb	Yes	Circular	—	12%	1 to 3	13/13	Peterson et al., 1993
tyrosinase	YAC	250 kb	Yes	Circular	12%	7%	1 to 8	5/5	Schedl et al., 1993b
tyrosinase	YAC	35 kb	Yes	Circular	24%	28%	1 to 50	5/9	Schdel et al., 1992

[a] If yes, then cloning vector sequences were present in DNA sequences microinjected for transgenic production. If no, then vector DNA was removed prior to microinjection.

[b] % Born is the proportion of microinjected eggs transferred to surrogate mothers that developed normally to birth.

[c] % Transgenic is the proportion of animals that developed from microinjected eggs that contained the transgene.

[d] Copy Number is the number of transgenes that integrated onto the mouse chromosome.

[e] Expressing Founders is the number of mice with independent integration events that expressed the gene contained in the transgene. Data are given only for animals that contained substantially intact transgenes.

[f] Three overlapping P1 clones were co-injected and underwent homologous recombination after injection to form a 180 kb transgene array.

without forming aggregates large enough to clog microinjection needles (Montoliu et al., 1995; Bauchwitz and Constantini, 1998).

High-quality DNA is important for successful production of transgenic mice. Purification of BAC or P1 DNA for pronuclear microinjection is carried out by alkaline lysis kits commercially available from Qiagen or Machery-Nagel (distributed by Clontech). DNA yields are lower than obtained from preparations of plasmid cloning vectors. YAC DNA is isolated by pulsed-field gel electrophoresis to purify the YAC from endogenous host chromosomes (Schedl et al., 1993; Bauchwitz and Constantini, 1998). In addition to the physical microinjection of YAC DNA into fertilized eggs, transgenic mice can be produced from mouse ES cells by spheroplast fusion or lipofection with YAC DNA (Peterson, 1997). The YAC in vivo complementation experiments described here relied on microinjection for mouse transgenesis (Smith et al., 1997; Madjumder et al., 1998).

We offer the following method for the purification of unsheared DNA from BAC or P1 clones for pronuclear microinjection. Updates on the method will be available at http://www.med.umich.edu/tamc/BACDNA.html. Prepare a 500-ml bacterial culture in LB, and harvest the bacteria when the O.D. 600 reaches 1.0. Lyse the bacterial pellet in 20 ml of buffer S1 (see Table 1.5 for buffer compositions). Add 20 ml of buffer S2. Do not vortex, but mix the tube

Table 1.4. Technical Differences in Pronuclear Injection Between Small DNA Inserts Cloned in Plasmid or Cosmids and Large DNA Fragments Cloned in YAC, BAC, or P1 Vectors

Plasmid or Cosmid Inserts	YAC, BAC, or P1 Inserts
Cloned DNA present in multiple copies per bacterial cell	Cloned DNA present in one or few copies per host cell
DNA is robust, resists breakage	DNA is fragile, easily broken
DNA resuspended in standard microinjection buffer[a]	DNA resuspended in polyamine microinjection buffer[b]
Gene expression reduced when cloning vector is present	No apparent effect of vector sequences on gene expression
Linearized DNA integrates more efficiently than circular DNA	Linearized and intact DNA integrate equally well
Equal proportion pups are transgenic with small or large DNAs	
Hundreds of DNA molecules are introduced by microinjection	As many as 10 DNA molecules are introduced by microinjection
Microinjection needle geometry:	Microinjection needle geometry:
long slender taper to shoulder	rapid taper to shoulder
Needle-tip diameter up to 1 μM	Needle-tip diameter up to 3 μM
Up to 50% of injected eggs develop normally	Up to 15% of injected eggs develop normally
A single genetic marker is sufficient for genotyping	Multiple markers are needed for accurate genotyping

[a] Standard microinjection buffer: 10-mM Tris, pH 7.4, 0.25-mM EDTA.

[b] Polyamine microinjection buffer: 10-mM Tris, pH 7.5, 0.1-mM EDTA, 100-mM NaCl, 30-μM spermine, 70-μM spermidine. Also see Table 1.3.

Table 1.5. Buffers for BAC and P1 DNA Purification

Buffer S1:	50 mM Tris, pH 8.0, 10 mM EDTA, pH 8.0 100 μg/ml DNAse free RNAse A	Store at 4° C
Buffer S2:	200-mM NaOH 1% SDS	Store at room temperature
Buffer S3:	2.8 M K-Acetate, pH 5.2	Store at 4° C
Buffer N2:	100 mM Tris-HCl pH 6.3 900 mM KCl 15% ethanol	Store at room temperature (adjust pH with H_3PO_4)
Buffer N3:	100 mM Tris-HCl pH 6.3 1150 mM KCl 15% ethanol	Store at room temperature (adjust pH with H_3PO_4)
Buffer N5:	100 mM Tris-HCl pH 8.5 1000 mM KCl 15% ethanol	Store at room temperature (adjust pH with H_3PO_4)
Polyamine Buffer:	10 mM Tris-HCl, pH 7.5 0.1 mM EDTA 100 mM NaCl 30 μM spermine tetrahydrochloride (Sigma cat. no. S-1141) 70 μM spermidine trihydrochloride (Sigma cat. no. S-2501)	

S1, S2, S3, N2, N3, N5 buffers, Nucleobond filters, and Nucleobond AX-500 cartridges are manufactured by Machery-Nagel and can be purchased from Clontech.

gently by inversion or by rolling the tube. Incubate at room temperature for 5 min. Add 20 ml of buffer S3 and gently invert 6 to 8 times until a homogenous mixture is produced. Incubate on ice for 10 min. Set up a 50-ml tube with a Nucleobond filter (moistened with 1 ml of sterile water) to remove the precipitated proteins and chromosomal DNA. Equilibrate a Nucleobond AX-500 column with 5 ml of buffer N2 and then run the filtered bacterial lysate through the column. The plasmid DNA will bind to the matrix. Wash the column twice with 20 ml of buffer N3. Elute the DNA from the column with 6 ml of buffer N5. Repeat the elution, and pool the eluates together. Precipitate the DNA with 0.7 volumes of room-temperature isopropyl alcohol. Pellet the DNA by centrifugation at 12,000 x g at 4°C for 15 minutes. Wash the pellet with cold 70% ethanol, air dry for about 5 min, and resuspend the pellet in 100 μl of microinjection buffer. Analysis of the BAC DNA by optical density, fluorometric measurements or standard gel electrophoresis is not uniformly effective. The best way to get a true picture of the quality of the DNA preparation is pulsed-field gel electrophoresis (Fig 1.3). We usually run two lanes of the DNA preparation, one with 1 μl, and one with 10 μl. DNA concentration is estimated by comparison of the intensity of ethidium bromide staining of the BAC DNA with that of known standards. The DNA concentration is adjusted to 1–3 ng/μl with polyamine buffer prior to microinjection.

There are other differences between transgenesis with small DNA inserts (plasmid, lambda, and cosmid cloning vectors) and the large (YAC, BAC, and P1 cloning vectors) listed in Table 1.2. Inclusion of vector sequences significantly reduces the expression of some transgenes relative to the level obtained in transgenic mice produced without vector DNA (Townes et al., 1985). The efficiency of producing transgenic mice is higher if the transgene is linearized

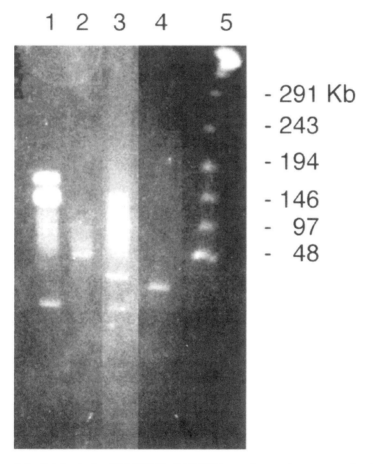

Figure 1.3. Gel Analysis of BAC DNA. BAC DNAs used to produce transgenic mice by pronuclear microinjection were analyzed by pulsed-field gel electrophoresis. Three different BAC DNA preparations digested with Not I. Lane 1: BAC 1086 (1-μl load), BAC 1097 (1-μl load), BAC 1080 (10-μl load). A 17-kb plasmid DNA standard (lane 4, ~15 ng DNA) and a lambda ladder molecular weight standard (lane 5, New England Biolabs) were used for comparison. The DNA in lane 1 is intact and highly concentrated, requiring dilution prior to microinjection. Variable amounts of bacterial chromosomal DNA detected in the pulse-field gel were not apparent in 0.8% agarose gels. The gel was run at 170 volts for 20 hours with a switch interval of 20 seconds.

prior to microinjection (Brinster et al., 1985). Neither of these observations appears to hold true for large DNA transgenes. Acceptable expression of transgenes associated with YAC, BAC, and P1 vector sequences has been reported, which suggests that the vector sequences are not problematic (Table 1.3). Whereas few direct comparisons are available, a tyrosinase minigene known to be expressed reliably in transgenic mice was equally efficient for expression when injected in the presence of YAC vector sequences (Schedl et al., 1992). Additional examples of YACs with good expression levels have been reported (Peterson, 1997). No significant difference in the efficiency of producing transgenics with either linearized or circular DNA has been reported (Table 1.3). In one study, linear and circular BACs were equivalent in their efficiency of generating transgenic mice, 6% and 9%, respectively (Antoch et al., 1997). This trend was confirmed independently with two BACs (Yang, 1998). A linear 131-kb BAC produced 12% transgenics, whereas the same circularized BAC yielded 7% transgenic; and a 112-kb BAC produced 11% and 17% transgenic with the linear and circular forms, respectively.

There are a number of biophysical differences between microinjection of large and small DNA species (Table 1.4). Typical experiments are designed to microinject a few hundred copies of the transgene at concentrations of 1 to 2 ng/μl. This arrangement gives the best efficiency of transgenesis and minimizes toxicity observed at higher DNA concentrations (Brinster et al., 1985). In experiments with large DNA molecules, DNA concentrations of 1 to 4 ng/μl have been used (Linton et al., 1993; Schedl et al., 1993; Smith et al., 1995; Antoch et al., 1997; Madjumder et al., 1998). This technique results in the transfer of many fewer DNA molecules into the pronucleus. Despite this condition, the percentage of transgenic pups from microinjections is comparable with percentages obtained from small DNA microinjection (Table 1.3). The number of transgene copies that integrate into the chromosome is generally lower than for small fragments, especially for transgenes 100 kb or greater in size. This result probably reflects the smaller number of microinjected molecules and is in agreement with the model proposed by Bishop and Smith (1989). They suggest that the head-to-tail concatamer of transgenes observed in the majority of integration sites is the result of homologous recombination between circularly permuted copies of the transgene. The number of available transgene copies is determined by the molarity of the transgene DNA injected, which is necessarily lower for large molecules such as BACs than small transgenes intended to express cDNA clones.

In addition to the constraints imposed by the size of large genomic inserts and the biology of their cloning vectors, there are practical considerations that influence the efficiency of producing transgenic mice with these molecules. For example, to prevent DNA aggregates from clogging injection needles during filling or injections, needle geometry can be modified so that the lumen of the needle widens rapidly from the tip to the shoulder. This modification usually makes the tip very brittle, which allows it to be broken open to a slightly larger diameter than usual. These two modifications reduce the number of needle changes necessary during a microinjection session. Although it is difficult to exactly reproduce micropipets between pipet puller settings on the basis of numbers, we offer the changes we made. We use a Sutter P-87 Flaming/Brown micropipette puller equipped with a 5-mm trough filament, air pump set to 190, and omega-dot glass capillaries (cat. no. TW100F-4 from World Precision Instruments). To obtain the desired needle shape we adjusted heat down to 390 from 450; kept pull parameter (softness that glass reaches prior to pulling apart capillary) at 100; reduced velocity from 70 to 60 (how hard the capillary is pulled apart); and increased time from 145 to 175 (how quickly filament is cooled after pull is initiated).

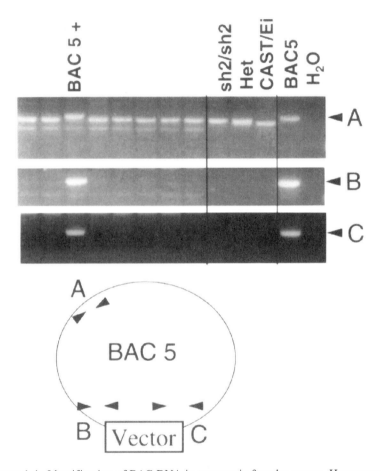

Figure 1.4. Identification of BAC DNA in transgenic founder mouse. Homozygous *shaker-2* mutant eggs were injected with BAC 5 DNA (140 kb). DNA was isolated from eight mice that were subsequently born. DNA was amplified with three PCR primer pairs internal to the BAC (shown at the bottom of the figure). A: primers around an internal Not I site. B: primer pair anchored in the 5' end of the cloning vector. C: primer pair anchored in the 3' end of the cloning vector. Agarose gel electrophoresis of PCR products showed that one of eight animals was transgenic and amplified with all three primer pairs (Lane BAC 5+). Primer pair A amplified the same product from seven nontransgenic animals (unmarked lanes) as from homozygous *shaker-2* mice (Lane sh2/sh2). Het: amplification target is DNA from offspring of *shaker-2* and CAST/Ei mouse mating. CAST/Ei: amplification target is CAST/Ei DNA. BAC 5: amplification target is purified BAC 5 DNA. H_2O: negative control showing lack of amplification in the absence of added DNA.

Low birth rates have been reported from YAC and BAC DNA microinjections (Table 1.3). When small DNA constructs are microinjected, birth rates are typically 20% (Brinster et al., 1985). The lower birth rates associated with high-molecular-weight DNA may be a consequence of DNA concentrations that are too high. It was noted that BAC DNA injected at 0.6 ng/μl generated transgenic pups, whereas a higher concentration (3 ng/μl) did not (Yang et al., 1997). The highest birth rate reported was 16% (Table 1.3). In our experience, DNA quality and concentration have the greatest impact on the number of pups born.

A critical element of any transgenic mouse project is the ability to detect the transgene and distinguish its expression from that of endogenous genes. Because transgenic mice produced with large DNA inserts from YAC, BAC, or P1 vectors may contain incomplete DNA fragments, it is valuable to use two or more assays to detect the transgene (Fig. 1.4). For example, PCR detection of the 5' and 3' ends and internal transgene marker will suggest that the entire construct has been integrated (Probst et al., 1998). If only one or two of the markers is detected, then a partial copy of the construct may have integrated. Such partial clones can be useful for delimiting the critical region of a clone in a complementation analysis. The integrity of the integrated DNA can be verified by restricting enzyme analysis and using Southern blotting.

8. Conclusion

The advantage of in vivo complementation with large genomic fragments is the rapid exclusion of large regions of the genetic map. This advantage rapidly narrows the critical region of the mutation and reduces the number of candidate genes that need to be analyzed for the mutation. In the ideal case, the genetic interval is narrowed to the single gene responsible for the phenotype. To be successful, careful attention to the identification of transgenic animals, the preparation of high-quality, intact microinjection DNA, and the map of the genetic locus under study are all necessary. Although YAC clones are important resources for genomic analysis, BAC clones might be better-suited for complementation studies by microinjection. Although YACs contain more genomic DNA than do BACs, there are often problems with instability and chimerism. Despite their smaller average insert size, many genes can be entirely contained in single BAC clones. Both BACs and YACs can be modified by genetic methods. Genetic complementation with large DNA molecules is a powerful application of established transgenic technology and more recent, still evolving, genomic cloning systems. The combination of these methods provides a potent tool for the identification and generation of animal models of human disease and disease genes. This knowledge can then be translated into treatments for human patients.

Acknowledgments

Thanks to Matt Callow, Amos Heckendorf, Rick Kaufman, Lluis Montoliu, Chris Russell, and X. William Yang for sharing DNA purification insights and methods. Thanks to Maggie Van Keuren, Frank Probst, Dave Kohrman, and Merle Rosenzweig for their assistance.

References

Antoch MP, Song EJ, Chang AM, Vitaterna MH, Zhao Y, Wilsbacher LD, et al. (1997): Functional identification of the mouse circadian *clock* gene by transgenic BAC rescue. Cell 89:655–667.

Barbosa MDFS, Nguyen QA, Tchernev VT, Ashley JA, Detter JC, Blaydes SM, et al. (1996): Identification of the homologous beige and Chediak-Higashi syndrome genes. Nature 382:262–265.

Bauchwitz R, Constantini F (1998): YAC transgenesis: a study of conditions to protect YAC DNA from breakage and a protocol for transfection. Bioc Biop Acta 1401:21–37.

Bedell MA, Largaespada DA, Jenkins NA, Copeland NG (1997): Mouse models of human disease. Part II: recent progress and future directions. Gene Dev 11:11–43.

Best S, le Tissie P, Towers G, Stoye JP (1996): Positional cloning of the mouse retrovirus restriction gene Fv1. Nature 382:826–829.

Bishop JO, Smith P (1989): Mechanism of chromosomal integration of microinjected DNA. Mol Biol Med 6:283–298.

Boehm T (1998): Positional cloning and gene identification. Methods 14:152–158.

Brinster RL, Chen HY, Trumbauer ME, Senear AW, Warren R, Palmiter RD (1985): Factors affecting the efficiency of introduction of foreign DNA into mice by microinjecting eggs. PNAS USA 82:4438–4442.

Callow MJ, Stoltzfus LJ, Lawn RM, Rubin EM (1994): Expression of human apolipoprotein B and assembly of lipoprotein(a) in transgenic mice. PNAS USA 91(6):2130–2134.

Camper SA (1985): Research applications of transgenic mice. Biotechniques 5:638–650.

Carstea ED, Morris JA, Coleman KG, Loftus ST, Zhang D, Cummings C, et al. (1997): Niemann-Pick C1 disease gene: homology to mediators of cholesterol homeostasis. Science 275:228–231.

Chiesa G, Johnson DF, Yao Z, Innerarity TL, Mahley RW, Young SG, et al. (1993): Expression of human apolipoprotein B100 in transgenic mice. Editing of human apolipoprotein B100 mRNA. J Biol Chem 268:23747–23750.

Dietrich WF, Copeland NG, Gilbert DJ, Miller JC, Jenkins NA, Lander ES (1995): Mapping the mouse genome: current status and future prospects. PNAS USA 92:10849–10853.

Gu JZ, Carstea ED, Cummings C, Morris JA, Loftus SK, Zhang D, et al. (1997): Substantial narrowing of the Niemann-Pick C candidate interval by yeast artificial chromosome complementation. PNAS USA 94(14):7378–83.

Haldi ML, Strickland C, Lim P, VanBerkel V, Chen X, Noya D, et al. (1996): A comprehensive large-insert yeast artificial chromosome library for physical mapping of the mouse genome. Mamm Genome 7:767–769.

Hamilton BA, Smith DJ, Mueller KL, Kerrebrock AW, Bronson RT, van Berkel V, et al. (1997): The vibrator mutation causes neurodegeneration via reduced expression of PITPalpha: Positional complementation cloning and extragenic suppression. Neuron 18:711–722.

Hauser MA, Amalfitano A, Kumar-Singh R, Hauschka SD, Chamberlain JS (1997): Improved adenoviral vectors for gene therapy of Duchenne muscular dystrophy. Neuromusc D 7:277–283.

Hogan B, Beddington R, Constantini F, Lacy E (1994): *Manipulating the Mouse Embryo: A Laboratory Manual*, Second Edition; Cold Spring Harbor Laboratory Press: Cold Spring Harbor, NY.

Ioannou PA, Amemiya CT, Garnes J, Kroisel PM, Shizuya H, Chen C, et al. (1994): A new bacteriophage P1-derived ector for the propagation of large human DNA fragments. Nat Genet 6:84–89.

Keegan CE, Karolyi IJ, Burrows HL, Camper SA, Seasholtz AF (1994): Homologus recombination in fertilized mouse eggs and assessment of heterologous locus control regions function. Transgenics 1:439–449.

King DP, Zhao Y, Sangoram AM, Wilsbacher LD, Tanaka M, Antoch MP, et al. (1997): Positional cloning of the mouse circadian clock gene. Cell 89:641–653.

Kling C, Koch M, Saul B, Becker CM (1997): The frameshift mutation oscillator (Glra1(spd–ot)) produces a complete loss of glycine receptor alpha1-polypeptide in mouse central nervous system. Neuroscience 78:411–417.

Liang Y, Wang A, Probst FJ, Arhya IN, Barber TD, Chen KS, et al. (1998): Genetic mapping refines DFNB3 to 17p11.2, suggests multiple alleles of DFNB3, and supports homology to the mouse model shaker-2. Am J Hu Gen 62:904–915.

Linton MF, Farese RV Jr., Chiesa G, Grass DS, Chin P, Hammer RE, et al. (1997): Transgenic mice expressing high plasma concentrations of human apolipoprotein B100 and lipoprotein(a). J Clin Inv 92:3029–3037.

Madjumder K, Shawlot W, Schuster G, Harrison W, Elder FFB, Overbeek PA (1998): YAC rescue of downless locus mutations in mice. Mamm Genome 9:863–868.

Manson AL, Trezise AEO, MacVinish LJ, Kasschau KD, Birchall N, Episkopou V, et al. (1997): Complementation of null CF mice with a human CFTR YAC transgene. EMBO J 16:4238–4249.

McCormick SP, Linton MF, Young SG (1994): Expression of P1 DNA in mammalian cells and transgenic mice. Genet A-Tech Appl 11:158–164.

McCormick SP, Nielsen LB (1998): Expression of large genomic clones in transgenic mice: new insights into apolipoprotein B structure, function, and regulation. Curr Op Lip 9:103–111.

Meisler MH (1996): The role of the laboratory mouse in the human genome project. Am J Hu Gen 59:764–771.

Montoliu L, Bock CT, Schutz G, Zentgraf H (1995): Visualization of large DNA molecules by electron microscopy with polyamines: application to the analysis of yeast endogenous and artificial chromosomes. J Mol Biol 246:486–492.

Nielsen LB, Kahn D, Duell T, Weier HU, Taylor S, Young SG (1998): Apolipoprotein B gene expression in a series of human apoliprotein B transgenic mice generated with recA assisted restriction endonuclease cleavage modified bacterial artificial chromosomes. J Biol Chem 273:21800–21807.

Palmiter RD, Brinster RL, Hammer RE, Trumbauer ME, Rosenfeld MG, Birnberg NC, et al. (1982): Dramatic growth of mice that develop from eggs microinjected with metallothionein-growth hormone fusion genes. Nature 300:611–615.

Paszty C, Brion CM, Manci E, Witkowska HE, Stevens ME, Mohandas N, et al. (1997): Transgenic knockout mice with exclusively human sickle hemoglobin and sickle-cell disease. Science 278:876–878.

Perou CM, Moore KJ, Nagle DL. Misumi DJ, Woolf EA, McGrail SH, et al. (1996): Identification of the murine beige gene by YAC complementation and positional cloning. Nat Genet 13:303–308.

Peterson KR. Production and analysis of transgenic mice containing yeast artificial chromosomes. In *Genetic Engineering*; Setlow JK, Ed.; Plenum Press: New York, 1997; 19, pp 235–255.

Probst FJ, Chen KS, Zhao Q, Wang A, Friedman TB, Lupski JR, et al. (1999): A physical map of the mouse shaker-2 region contains many of the genes commonly deleted in Smith-Magenis syndrome. Genomics 55:348–352.

Probst FJ, Fridell RA, Raphael Y, Saunders TL, Wang A, Liang Y, et al. (1998): Correction of deafness in shaker-2 mice by an unconventional myosin in a BAC transgene. Science 280:1444–1447.

Rozmahel R, Gyomorey K, Plyte S, Nguyen V, Wilschanski M, Durie P, et al. (1997): Incomplete rescue of cystic fibrosis transmembrane conductance regulator deficient mice by the human CFTR DNA. Hum Mol Gen 6:1153–1162.

Schedl A, Beermann F, Thies E, Montoliu L, Kelsey G, Schutz G (1992): Transgenic mice generated by pronuclear injection of yeast artificial chromosome. Nucl Acid R 20:3073–3077.

Schedl A, Larin Z, Montoliu L, Thies G, Kelsey G, Lehrach H, Schutz G (1993): A method for the generation of YAC transgenic mice by pronuclear microinjection. Nucl Acid R 21:4783–4787.

Schimenti J, Bucan M (1998): Functional genomics in the mouse: phenotype-based mutagenesis screens. Genome Res 8:698–710.

Scott J, Wallis SC, Davies MS, Wynne JK, Powell LM, Driscoll DM (1989): RNA editing: a novel mechanism for regulating lipid transport from the intestine. Gut 30 (Suppl 1):35–43.

Shawlot W, Siciliano MJ, Stallings RL, Overbeek PA. (1989): Insertional inactivation of the downless gene in a family of transgenic mice. Mol Biol Med 6:299–307.

Smith DJ, Stevens ME, Sudanagunta SP, Bronson RT, Makhinson M, Watabe AM, et al. (1997): Functional screening of 2 mb of human chromosome 21q22.2 in transgenic mice implicates minibrain in learning defects associated with Down's syndrome. Nat Genet 16:28–36.

Smith DJ, Zhu Yiwen, Zhang JL, Cheng JF, Rubin EM (1995): Construction of a panel of transgenic mice containing a contiguous 2-mb set of YAC/P1 clones from human chromosome 21q22.2. Genomics 27:425–434.

Townes TM, Lingrel JB, Chen HY, Brinster RL, Palmiter RD (1985): Erythroid-specific expression of human beta-globin genes in transgenic mice. EMBO J 4:1715–1723.

Veniant MM, Kim E, McCormick S, Boren J, Nielsen LB, Raabe M, et al. (1999): Insights into apolipoprotein B biology from transgenic and gene-targeted mice. J Nutr 129:451S–455S.

Wagner SD, Gross G, Cook GP, Davies SL, Neuberger MS (1996): Antibody expression from the core region of the human IgH locus reconstructed in transgenic mice using bacteriophage P1 clones, MRC Laboratory of Molecular Biology, Cambridge, United Kingdom. Genomics 35(3):405–415.

Wakabayashi Y, Kikkawa Y, Matsumoto Y, Shinbo T, Kosugi S, Chou D, et al. (1997): Genetic and physical deliniation of the region of the mouse deafness mutation shaker-2. Bioc Biop R 234:107–110.

Wang A, Liang Y, Fridell RA, Probst FJ, Wilcox ER, Touchman JW, et al. (1998): Association of Unconventional Myosin MYO15 mutations with human nonsyndromic deafness DFNB3. Science 280:1447–1451.

Xiong W, Zsigmond E, Gotto AM Jr., Reneker LW, Chan L (1992): Transgenic mice expressing full-length human apolipoprotein B-100. Full-length human apolipoprotein B mRNA is essentially not edited in mouse intestine or liver. J Biol Chem 267:21412–21420 [retracted in J Biol Chem 1993; 268:17647].

Xiong WJ, Zsigmond E, Gotto AM Jr., Lei KY, Chan L (1991): Locating a low-density lipoprotein-targeting domain of human apolipoprotein B-100 by expressing a minigene construct in transgenic mice. J Biol Chem 266:20893–20898 [Retracted in J Biol Chem 1993; 268:17647].

Yang XW (1998): Genetic control of cerebellar granule cell production and cerebellar morphogeneis by murine zinc finger transcription factor RU49. Dissertation.

Yang XW, Model P, Heintz N (1997): Homologous recombination-based modification in *Eschirichia coli* and germline transmission in transgenic mice of a bacterial artificial chromosome. Nat Biotech 15:859–865.

Zhang Y, Buchholz F, Muyrers JP, Stewart AF (1998): A new logic for DNA engineering using recombination in *Escherichia coli*. Nat Genet 20: 123–128.

Zuberi AR, Christianson GJ, Sonal BD, Bradley JA, Roopenian DC (1998): Expression screening of a yeast artificial contig refines the location of the mouse H3a minor histocompatibility antigen gene. J Immunol 161:821–828.

CHAPTER 2

BEHAVIORAL CHARACTERIZATION OF TRANSGENIC AND KNOCKOUT MICE

JACQUELINE N. CRAWLEY, PH.D.

1. Introduction

Targeted gene mutation provides a powerful tool for investigating the role of a specific receptor subtype in mediating the behavioral actions of a neurotransmitter. This genetic approach is particularly useful when no subtype-selective pharmacological tools are available. Gene deletions for each of the receptor subtypes can be used to determine which behavioral function is mediated by which receptor subtype. If administration of the naturally occurring receptor agonist fails to produce an expected behavioral action, then the mutated receptor subtype is likely to mediate that behavioral effect. This chapter will describe behavioral methods to analyze phenotypes of transgenic and knockout mice, with emphasis on mutations in genes for neurotransmitter receptor subtypes.

2. Experimental Design

Several factors must be considered in designing the behavioral phenotyping experiments. The first is breeding strategy. Most mutations are generated in embryonic stem cells originating from one of the 129 inbred substrains of mice. Unfortunately, several of the 129 substrains contain background genes that produce aberrant behavioral phenotypes. For example, 129/J mice fail to develop the corpus callosum fiber bundle that connects the right and left cerebral cortex hemispheres, and show severe deficits in learning and memory tasks (Livy and Wahlsten, 1997; Montkowski et al., 1997). Further, most blastula donation and breeding is conducted in a strain that produces large numbers of offspring.

Genetic Manipulation of Receptor Expression and Function,
Edited by Domenico Accili.
ISBN 0-471-35057-5   Copyright © 2000 Wiley-Liss, Inc.

However, good breeding strains such as CD-1 are outbred, which dramatically increases the variability within the control group, thereby requiring a much bigger effect of the mutation to be detectable above the baseline variability of the controls.

The best approach is to breed a set of mutant mice specifically for behavioral phenotyping by backcrossing into an inbred strain with characteristics appropriate for the hypothesized functions of the gene of interest. Characteristics of a variety of inbred strains of mice, on a wide range of behavioral tests, have been published in several good reviews of mouse strain distributions (Wehner and Silva, 1996; Crawley et al., 1997; Banbury Conference, 1997). Choosing the optimal background strain for breeding will avoid complications of background genes that mask the behavioral phenotype of the mutation, due to "ceiling" or "floor" effects. For example, a strain with high levels of aggressive behaviors is useful if the gene to be mutated is likely to decrease aggression. A strain with low levels of locomotor activity is useful if the gene to be mutated is predicted to increase locomotion. A strain such as C57BL/6J shows moderate scores on most behavioral tests, making it a reasonable choice in many cases. Once the breeding strain is chosen, backcrossing for at least seven generations is required to produce a relatively pure genetic background (Silver, 1995). Five generations may be sufficient if using the speed congenic breeding strategy, wherein breeding males are selected by genetic markers and their scores on the relevant behavioral tests (Markel et al., 1997).

Genotypes and group size are generated from the final breeding generation. Homozygous (null) mutants ($-/-$) and heterozygous mutants ($+/-$) are compared with homozygous wild-type littermate controls ($+/+$). Environmental factors such as parental care, birth order, and dominance hierarchy in the home cage have major effects on many behavioral traits in mice. Use of littermates will control for these variables. Group size is a minimum of 10 mice per treatment group to achieve sufficient power for the standard statistical tests used in behavioral neuroscience. The treatment groups are males and females of each of the three genotypes. Thus, at least 60 mice are tested in the first experiments. If no gender differences are detected, the data from males and females can be combined for each genotype to increase the power of the statistical analysis. Gender data can then be pooled in future replication experiments, thereby reducing the required Ns.

If it is impractical to obtain the full complement of mice all at once, a series of smaller group sizes can be tested. The critical design feature is inclusion of some individuals from all three genotypes and both genders within each day's experiments. Data from the groups tested at different times can subsequently be combined for statistical analysis of genotype effects, unless a significant difference is detected between the time points.

Age may be a critical factor for some genes. For example, amyloid precursor protein overexpressing transgenic mouse models of Alzheimer's disease may express amyloid plaques, neurodegeneration, and memory impairment only at later ages (Holcomb et al., 1998). Most behavioral tests can be repeated at defined ages. Mice are considered "juvenile" at 2–6 weeks, "sexually mature adults" at 3–8 months, and "aged" at 1–2 years.

Genotyping is conducted on all mice. The molecular geneticist determines that the mutation is expressed as expected, and that the gene product is overexpressed in the transgenics or missing in the knockouts, in the generation of mice to be used for behavioral phenotyping. Confirmation of expected anatomical and biochemical sequelae of the mutation is best performed before the start of the labor-intensive behavioral testing.

Identification of individual mice is critical for behavioral experiments, because each animal is used for multiple behavioral tests. Firmly attached ear tags and/or subcutaneous barcode chips work well. Housing in the vivarium requires separate cages for males and females. Mice can be housed by genotype or housed as mixed genotypes. For behavioral experiments, the animals live in a quiet, temperature-controlled, humidity-controlled environment, on a fixed daily lighting schedule, and they are not simultaneously used for any other experiments.

3. Preliminary Observations: How to Avoid False Positives

Our laboratory developed a series of general observations for the initial evaluation of a new set of mutant mice. The goal is to identify gross physiological abnormalities. A sick mouse will perform poorly on all behavioral tasks. A blind mouse cannot learn a visual task; a deaf mouse will fail on acoustic startle tests; motor dysfunctions will impair performance on every behavioral task that requires movement. Recognizing gross functional abnormalities is particularly important for conventional mutations. When the mutation is expressed in all tissues of the body, from the earliest stages of development, peripheral organs may be impaired in ways that dramatically affect behaviors. Technological advances in conditional, inducible mutations will avoid some of these complications by expressing the mutation only in the brain region of interest and only during the experimental time period.

False positives are avoided by recognizing the limitations of the mouse and designing specific behavioral tasks around the limitations. For example, learning and memory tests that use olfactory and auditory cues can be used to test mice with visual impairment. True positives are often revealed in the course of the general observations. Mouse models of human ataxia and motor degeneration are well characterized by behavioral tests for locomotor activity, balance, and coordination. For example, a mouse model of Tay-Sachs and Sandhoff diseases that contains a mutation in hexosaminidase, the enzyme that degrades gangliosides, was generated by Rick Proia and coworkers at NIDDK. *Hexb* mutant mice show a progressive loss of ability to perform on the rotarod task (Sango et al., 1995), analogous to the motor deficits that characterize the human syndrome. In a mouse model of Ataxia Telangiectasia, mutation of the *atm* gene generated mice with unusual footprint patterns (Barlow et al., 1996), analogous to the ataxia seen in the clinical syndrome. Dopamine transporter knockouts demonstrate hyperlocomotion in a photocell-equipped automated open field, analogous to the hyperlocomotion induced by treatment with drugs that increase synaptic dopamine levels (Giros et al., 1996).

Mice are first observed in their home cages. Overall health and condition of the fur and whiskers are noted. Home cage activity, grooming, nesting, and sleeping patterns are observed. Any unusual patterns of locomotion, hyperre-activity to handling, or fighting in the home cage are noted. Body weight is recorded by weighing each mouse in a standard triple-beam or pan balance. Body temperature is measured by rectal thermistor. Abnormal appearance, barbered whiskers, unusual body weight, and unusual home cage social behaviors can provide important clues for further experiments to define the behavioral phenotype. For example, high levels of aggressive behavior were first detected in the home cage in nitric oxide synthase knockout mice (Nelson et al., 1995). Unusual nesting patterns led to the discovery of social interaction abnormalities in dishevelled-1 knockout mice (Lijam et al., 1997).

4. Neurological Reflexes, Motor Skills, and Sensory Abilities

Batteries of reflexes have been described by Irwin (1968), Moser (1995), and the SHIRPA team (Rogers et al., 1997). Richard Paylor recently adapted elements of the Irwin screen for use in our laboratory, shown in Table 2.1 as a checklist (Crawley and Paylor, 1997). The righting reflex test measures the time it takes for the mouse to right itself to an upright posture after being turned on its back. The eye blink reflex is elicited by approaching the eye with a cotton-tip swab. The ear twitch reflex is elicited by touching the ear with a cotton-tip swab. The whisker-orienting reflex is measured by touching the vibrissae on one side; the whiskers will normally stop moving and the head will turn to the side on which the whiskers were touched. Visual cliff behavior is rapidly evaluated by observing whether the mouse walks off the edge of a high table. Acoustic startle is rapidly evaluated by observing the flinch response to a sudden loud sound. Most mice are normal on all of the tests for simple neurological reflexes. If a reflex is abnormal, then the mutation has produced a gross physiological or behavioral phenotype that can be fruitfully investigated with more sophisticated tests.

Motor skills underlie the performance of almost all behavioral tasks. If the mutation results in severe deficits in the ability of the mouse to walk, grip, balance, climb, swim, and so on, the mouse may be unable to perform the procedures necessary for more sophisticated behavioral tasks. Several good paradigms are routinely used to evaluate motor functions. Open field exploratory activity is quantitated in a photocell-equipped automated Digiscan apparatus that records several parameters of locomotion and rearings. The ability of the mouse to balance on a rotating cylinder is quantitated with an automated accelerating rotorod apparatus. These two tests are sensitive to major abnormalities in spinal motor neurons and cerebellum. Grip tests, such as hanging wire grip time, measure neuromuscular strength. Footprint pathway analysis detects abnormal gait.

Specific sensory abilities are necessary for the performance of many behavioral tasks. Sensitive sensory tests for mice involve neurophysiological recording during presentation of the sensory stimulus. The auditory brainstem response is a neurophysiological measurement of the auditory nerve activity in response to

Table 2.1. Rapid Tests to Detect Gross Abnormalities in General Health, Neurological Reflexes, Sensory Abilities, and Motor Functions. (Adapted from Paylor et al., 1998.)

The Paylor Battery: General Health, Neurological Reflexes, Sensory and Motor Functions

A. Physical Characteristics

Body weight
Body temperature
Whiskers (% with)
Bald patches (% with)
Palpebral closure (% with)
Exophthalmos (% with)
Piloerection (% with)

B. General Behavioral Observations (% of mice displaying the response in a novel, bare-cage environment)

Wild running
Freezing
Sniffing
Licking
Rearing
Jumping
Defecating
Urinating
Moving around entire cage

C. Sensorimotor Reflexes (% mice displaying normal response)

Righting reflex
Whisker twitch
Eye blink
Ear twitch
Hot-plate latency to first hind-paw lick

D. Sensory and Motor Functions

Wire suspension test (latency to fall)
Pole test score
Tail suspension test
Elevated platform latency to edge
Exploratory nose pokes

---

a series of tones. The electroretinogram measures the optic nerve activity in response to grades flashes of light. Visual response to light is determined by measuring constriction and subsequent dilation of the pupil when a small flashlight beam is directed at the eye. Neurophysiological recording from olfactory nerve and olfactory cortex is the most sensitive measure of sensitivity to smell.

Sensitive measures of some sensory abilities can be conducted with complex behavioral tasks. Visual acuity is measured by responses to visual stimuli after mice have been trained in a conditioned reward paradigm. Acoustic threshold is evaluated in an automated startle system that measures amplitude of whole body flinch across a range of tones. Pain sensitivity is evaluated by a threshold determination, using jumps, vocalizations, and running in response to a graded series of foot shocks, or with the graded diameters of Von Frey hairs touched to the base of a foot. Olfactory ability is evaluated in a choice test for a series of odors delivered through a specialized airflow system, in a conditioned reward paradigm. Taste discriminations and preference are measured in a choice test with graded gustatory stimuli. All of these tests require specialized equipment and multiple training sessions. Simpler methods for evaluating sensory abilities in mice would be of great benefit to the behavioral phenotyping armamentarium.

5. Behavioral Domains: How to Avoid False Negatives

Behavioral neuroscience has a rich literature encompassing many well-established, carefully validated tests for mouse behaviors over a wide variety of domains. Several excellent reviews (Wehner and Silva, 1996; Campbell and Gold, 1996; Crawley et al., 1997, 1999; Crawley and Paylor, 1997; Jucker and Ingram, 1997; Nelson and Young, 1997) cite the primary literature on the best tests to evaluate learning and memory, feeding, sexual behaviors, parenting behaviors, social interactions, aggression, anxiety-like behaviors, depression-like behaviors, addictive behaviors, and schizophrenia-like symptoms, as well as vision, hearing, smell, taste, pain threshold, locomotion, balance, muscle strength, ataxia, and seizures. Our laboratory generally selects two or three tests from each behavioral domain relevant to the gene of interest. Each test measures a slightly different behavior, with different underlying mechanisms, and requires different sensory and motor modalities. The likelihood of true positives—detecting a specific behavioral abnormality revealing the function of the gene—is increased by conducting multiple, complementary tests within a given domain. The likelihood of false negatives is much higher when only a single test is used. If the same behavioral abnormality is detected across two or three complementary tests, the interpretation of the behavioral phenotype is very strong. For example, an anxiety-like phenotype on the light ↔ dark exploration task, the elevated plus maze, and the fear-conditioned startle corroborate the interpretation of increased fear or anxiety in a mutant line. If the behavioral abnormality is detected in only one of the complementary tests, the specific type of abnormality can be further explored in additional tests that focus on that component. For example, if cued and contextual conditioning is impaired but the Morris water task and radial maze acquisition are normal, future research will focus on fear-related learning tasks.

Elegant studies of specific behavioral domains in knockout mice deficient in receptor subtypes have greatly increased the understanding of the functional role of a given receptor subtype. Selected publications are described below. These examples are chosen to illustrate some of the best available tests for mice, across a wide range of behavioral domains. Similar methods can be effectively applied to investigate the behavioral phenotype of any receptor mutation.

6. μ Opiate-Receptor Subtype Null Mutants

Comprehensive, well-replicated experiments have been conducted with mice deficient in the μ-opiate-receptor subtype (reviewed in Kieffer, 1999). Null mutants deficient in the μ-receptor were generated in three independent laboratories: Brigitte Kieffer and coworkers, CNRS, Universite Louis Pasteur, Strasbourg, France; George Uhl and coworkers, National Institute on Drug Abuse, Baltimore, Maryland, USA; and Horace Loh and coworkers, University of Minnesota, Minneapolis, USA. Behavioral domains representing the major pharmacological actions of morphine were tested, including analgesia and reward.

Baseline levels of analgesia on the tail-flick and hot-plate tests were significantly lower in the μ-receptor homozygote null mutant mice as compared with wild-type littermate controls on latency to tail-flick, and on latency to paw-lick in the hot-plate test at $55°C$, in one report (Sora et al., 1997). Baseline levels of analgesia were not significantly different on latency to tail withdrawal, the tail immersion test, nor on measures of paw licking and jumping in the hot-plate test, in another report (Matthes et al., 1996). These discrepancies may reflect the contributions of different background genes from the different embryonic stem cell lines used to generate the mutation, and/or differences in the methods used for the behavioral tests, between the two laboratories. Baseline data on analgesia were not reported from the third laboratory (Loh et al., 1998).

Morphine-induced analgesia was abolished in all three μ-receptor knockout lines, as measured with the tail-flick and the hot-plate tests for pain sensitivity (Matthes et al., 1996; Sora et al., 1997; Loh et al., 1998). Further, analgesic responses to heroin and to a morphine metabolite, morphine-6-glucuronide, were absent in μ-receptor knockout mice (Kitanaka et al., 1998; Loh et al., 1998). Robust and consistent differences on a drug response, but minor or inconsistent baseline responses, are often the case with a receptor subtype mutation. Failure to show the expected difference in baseline behaviors may reflect compensation by another receptor subtype for that neurotransmitter, or may disprove the hypothesized function of the receptor subtype.

The role of the μ-opiate-receptor subtype in the reinforcing actions of morphine was explored in one of the μ-receptor mutant lines. Time spent in a chamber where mice previously received morphine is compared with time spent in an adjoining chamber where mice previously received saline vehicle. Conditioned place preference to morphine was demonstrated in the wild-type control mice, as previously reported for rats. Conditioned place preference to morphine was completely absent in the μ-receptor null mutants (Matthes et al., 1996).

Withdrawal symptoms after cessation of chronic treatment with morphine were prominent in the wild-type control mice, including wet-dog shakes, jumping, sniffing, teeth-chattering, tremor, and diarrhea. Withdrawal symptoms were absent in the μ-receptor knockout mice (Matthes et al, 1996). Morphine-induced lethality due to tonic/clonic convulsions was prominent at the dose of 800 mg/kg in the wild-type and heterozygotes but absent in the knockouts at that dose of morphine (Loh et al., 1998). Immune-suppressant effects of morphine, including lymph organ atrophy, reduced natural killer-cell activity, and diminished ratio of CD4+CD8+ cells in the thymus were absent in μ-receptor mutants (Gavériaux-Ruff et al., 1998).

Quantitative autoradiography confirmed the absence of μ-receptors, whereas no significant differences were detected in concentrations or localization of δ- or μ-opiate receptors, indicating no compensatory increases in other opiate-receptor subtypes in the μ-receptor knockout mice (Kitchen et al., 1997). Binding affinities and agonist-induced adenylyl cyclase inhibition indicated normal functional coupling of δ- and κ-receptors to G-proteins in brain and spinal cord tissue from the μ-receptor mutants (Matthes et al., 1998).

These complementary studies confirm a strong functional role of the μ-opiate-receptor subtype in the analgesic, rewarding, and immune-suppressant actions of morphine. Similar analyses are presently being conducted on δ- and κ-opiate-receptor subtype knockout mice (Keiffer, 1999). This approach has the power to test hypotheses about selective functions of opiate-receptor subtypes, evaluate the subtype selectivity of ligands, and understand the contributions of endogenous opiates at each receptor subtype to behavioral and physiological functions.

7. Dopamine-Receptor Subtype Mutations

Dopamine is a catecholamine neurotransmitter with multiple receptor subtypes (Sokoloff and Schwartz, 1995). Ligands of dopamine receptors have well-documented behavioral actions when microinjected into brain regions of rats, including actions on locomotor activity, grooming, feeding, prepulse inhibition, sexual behaviors, and responses to amphetamine and cocaine. Dopaminergic neurotransmission is implicated in neuropsychiatric disorders, including Parkinson's disease, schizophrenia, attention deficit hyperactivity disorder, and drug abuse. At least five dopamine-receptor subtypes have been identified to date. Targeted gene mutations of each have been generated and are being evaluated for their behavioral phenotypes. Analyses provide examples of tests used in the behavioral domains of locomotion, grooming, stereotypy, motor coordination, learning and memory, and responses to psychostimulants.

D1-deficient mice were not significantly different from wild-type littermate controls on open field locomotion, but failed to show the standard hyperlocomotion and stereotypy response to treatment with cocaine or with a D1 agonist, SKF 81297 (Xu et al., 1994a,1994b). D1-knockout mice displayed reduced grooming in response to novelty and to intraventricularly administered oxytocin, prolactin, ACTH, and β-endorphin (Drago et al., 1999). In contrast, D2-knockout mice displayed normal grooming and normal responses to novelty and neuropeptide administration (Drago et al., 1999). D_{1A}-deficient mice were not significantly different from wild-type littermates on an olfactory discrimination, on distance traversed and time spent in three zones of a circular open field, or on swim speed and pattern in the Morris water task. However, a significant impairment in place learning performance was detected in both the visible and hidden platform components of the Morris water task (Smith et al., 1998a). D2-deficient mice displayed reduced locomotion in the open field and poor performance on the rotarod (Balk et al., 1995). In addition, D2- knockout mice failed to show conditioned place preference to morphine, indicating loss of rewarding effects of opiates (Maldonado et al., 1997). D3-deficient mice showed higher levels of locomotion in an open field (Accili et al., 1996). D4-deficient mice displayed reduced baseline horizontal activity and locomotion in an open field but increased locomotor activation in response to ethanol, cocaine, and methamphetamine administration (Rubenstein et al., 1997). D5-deficient mice, currently being tested, show preliminary indications of hyperlocomotion in the Digiscan open field (Sibley et al., 1998).

As indicated by this brief summary, each laboratory generating a new dopamine- receptor subtype knockout mouse is presently conducting an independent and often unrelated set of behavioral tests. The power of the technology will be more fully realized when identical tests are conducted in each laboratory for all of the receptor subtype mutants, to compare results within and across laboratories. This testing will allow definitive conclusions to be drawn about the role of each receptor subtype in behavioral functions such as exploratory locomotion and the rewarding properties of drugs of abuse. Further, double knockouts can be bred to test the effects of deleting two pharmacologically similar receptor subtypes, such as D2 and D4, or two pharmacologically different subtypes, such as D1 and D2, to work out the complementary and opposing actions of each subtype on each behavioral function.

8. Corticotropin Releasing Factor Mutations

Corticotropin releasing factor (CRF) is a peptide neurotransmitter that is released from hypothalamic neurons in response to stressors (Behan et al., 1996). Two receptor subtypes for CRF have been identified to date. CRFR1 knockout mice were generated by Wylie Vale and coworkers at the Salk Institute and tested by Lisa Gold and coworkers at the Scripps Research Institute in La Jolla, CA (Smith et al., 1998b). This publication and an independent replication by Holsboer and coworkers at the Max Planck Institute of Psychiatry in Munich (Timpl et al., 1998) provide examples of good behavioral tests in the domains of anxiety and stress, as well as stress hormones in the hypothalamic-pituitary-adrenal axis. The elevated plus maze and the dark-light emergence test were used to evaluate anxiety-related behaviors. Significantly less anxiety-like behavior was found in the CRFR1 knockout mice on both tasks. Restraint stress induced large increases in circulating ACTH in the wild-type control mice but minimal increase in plasma ACTH in the CRFR1 knockout mice. The anxiolytic-like phenotype of CRFR1 knockout mice is consistent with previous publications on anxiogenic-like effects of CRF administration in rats (Sutton et al., 1982; Britton et al., 1986), anxiogenic-like phenotype of CRF overexpressing transgenic mice (Stenzel-Poore et al., 1994; Heinrichs et al., 1997), anxiolytic-like effects of CRF antagonists (Heinrichs et al., 1992; Lundkvist et al., 1996), and anxiolytic-like effects of antisense oligonucleotides (Skutella et al.,

1994), using the same and additional behavioral tests. CRF-receptor knockout mice are a useful research tool to confirm findings with selective-receptor ligands, to identify new subtype-selective ligands, and to evaluate hypotheses about receptor subtype selectivity for the actions of CRF on anxiety, stress responses, feeding, learning, sexual behaviors, and depression.

9. Neuropeptide Y-Receptor Subtype Mutations

Neuropeptide Y (NPY) is a widely distributed neurotransmitter with many biological actions, including blood-pressure regulation, seizures, anxiety-like behaviors, and induction of feeding (reviewed in Gehlert, 1998). At least six subtypes of NPY receptors have been cloned (reviewed in Inui, 1999). Pharmacological tools are being developed to determine which of the NPY-receptor subtypes mediates the ability of centrally administered NPY to stimulate high levels of food consumption in satiated rats. Simultaneously, Richard Palmiter and coworkers at the University of Washington in Seattle and Kushi and coworkers in Japan are generating mice deficient in NPY and NPY-receptor subtypes. Several methods have been applied to evaluate NPY and NPY-receptor knockout mice on feeding: 24-hr and weekly consumption of chow in the home cage, 2-hr consumption of chow after a one-day fast, consumption of a glucose solution, and daily body weights (Erickson et al., 1996a,b; Hollopeter et al., 1998; Kushi et al., 1998). Other methods to measure feeding, including limited daily access, high-fat and high-carbohydrate diets, and microstructural analysis of licking behavior, are described for evaluating other mutant mouse models of obesity and in the larger literature on feeding behaviors (reviewed in Hoebel and Wellman, 1998). NPY knockout mice showed normal body weights and normal daily consumption of chow (Erickson et al., 1996a), supporting the concept that endogenous NPY is one of many multiple, redundant neurotransmitters contributing to the regulation of appetite. Double mutants were then generated in which NPY knockout mice were bred with the natural mutant, *ob/ob*, an obese line deficient in the ob gene product, leptin. NPY deficiency partially blocked the development of obesity in adult *ob/ob* mice, indicating that NPY is required for the full manifestation of the obesity syndrome resulting from leptin deficiency in mice (Erickson et al., 1996b). Further, NPY-deficient mice were more sensitive to the inhibitory actions of leptin treatment on food consumption, as compared with wild-type controls (Hollopeter et al., 1998).

Both the NPY-Y1 and the NPY-Y5-receptor subtypes appear to contribute to the actions of NPY on feeding, as indicated by responses to relatively subtype-selective ligands (Gerald et al., 1996; Inui, 1999). NPY-Y1 knockout mice showed significantly higher body weight in the females, as compared with wild-type controls (Kushi et al., 1998). Food consumption, measured in the home cage on a weekly basis, appeared to be normal. More sensitive behavioral measures and/or pharmacological challenges may be required to detect differences in feeding behaviors in these NPY-Y1 knockout mice. NPY-Y5-receptor knockout mice developed late-onset obesity (Marsh et al., 1998). Body weight, feeding, and fasting-induced feeding were normal in young NPY-Y5 knockout mice, but older NPY-Y5 knockout mice were characterized by higher body weights, increased food consumption, and adiposity, as compared with wild-type littermate controls. Reduced or absent response to intraventricularly administered NPY on food consumption was demonstrated in the NPY-Y5 knockout mice. These data support an interpretation that the NPY-Y5-receptor mediates NPY-induced feeding, however, in the direction opposite to predicted. NPY-receptor subtype knockout mice provide a good tool for evaluating NPY-receptor selective agonists, to confirm that the agonist-stimulated feeding is mediated by the predicted receptor subtype. Single and double knockouts for each of the NPY-receptor subtypes will allow basic researchers to explicate the contribution of each of the NPY-receptor subtypes to the regulation of normal and abnormal ingestive behaviors.

10. Estrogen-Receptor Subtype Mutations

Estrogen is a gonadal steroid with direct effects on the transcription of genes in the nucleus of many cell types, including neurons regulating reproductive behaviors (Ramirez, 1992). Sexual behaviors in rodents are modulated by activation of estrogen receptors (Miesel and Sachs, 1994). Two estrogen receptors have been cloned, designed ERα and ERβ. ERα knockout mice were generated by Dennis Lubahn and coworkers at the University of Massachusetts in Amherst, and were tested with standard methods to evaluate male and female sexual behaviors by Emilie Rissman and coworkers at the University of Virginia in Charlottesville and Donald Pfaff and coworkers at Rockefeller University in New York (Rissman et al., 1997; Wersinger et al., 1997; Ogawa et al., 1997). Results provide examples of tests to measure female sexual behaviors through quantitation of lordosis behavior, and to measure male sexual behaviors through quantitation of mounts, intromissions, and ejaculations.

Female sexual receptivity was absent in female ERα knockout mice, following standard ovarectomy and treatment with either estradiol or a combination of estradiol and progesterone (Rissman et al., 1997). Male ERα similarly demonstrated reduced sexual behaviors, documented by longer latencies to mount and reduced frequency of mounts, thrusts, intromissions, and ejaculations (Wersinger et al., 1997). ERα knockout mice, and ERβ knockout mice when available, will be useful to evaluate receptor-subtype selectivity for other actions of estrogen in the brain, including its role in feeding and in memory.

11. Conclusions

The examples described above demonstrate the power of the targeted gene mutation technique for confirming and extending knowledge on the role of a particular receptor subtype in a specific behavioral function. In some cases, the mutant mouse approach simply confirmed the selectivity of receptor ligands on the behavioral function. In some cases, the mutant mouse approach confirmed the role of the receptor subtype in the behavioral function. In the most interesting cases, the behavioral phenotype detected new functions for the receptor subtype and its neurotransmitter.

In some cases, no significant behavioral phenotype was detected in the mutant line, in contrast to predictions based on pharmacological studies with receptor ligands in normal rats and mice. Lack of significant phenotype may disprove the hypotheses, or may be due to the technical limitations of the present targeted mutation approach. Redundant genes may compensate, or even overcompensate, for the mutation, which is present from the early blastocyst stage of development. Other physiological and behavioral processes may take over the function of the missing gene product over the course of development. Temporally inducible mutations, and tissue-specific conditional mutations, are necessary to avoid these limitations and increase the power of the mutant mouse technology.

The major advantage of the present conventional mutation approach is seen in cases where multiple receptor subtypes have been cloned but selective ligands have not been generated. For example, three subtypes of the galanin receptor are known, but there are not yet any highly selective peptidergic or nonpeptidergic ligands (Hökfelt et al., 1998). The null mutant mouse deficient in one galanin-receptor subtype may reveal the functional role of that subtype in a behavior such as memory, feeding, or aggression. The mutant line can further be used to evaluate the functional activity of ligands for other receptor subtypes in each relevant behavioral domain. Finally, the receptor subtype-selective knockout mouse can serve as an important tool for discovering new therapeutic treatments for neuropsychiatric diseases.

References

Accili D, Fishburn CS, Drago J, Steiner H, Lachowicz J, Park RH, Gauda EB, Lee EJ, Cool MH, Sibley DR, Gerfen CR, Westphal H, Fuchs S (1996): A targeted mutation of the D_3 dopamine receptor gene is associated with hyperactivity in mice. PNAS USA 93:1945–1949.

Balk JH, Picetti R, Salardi A, Thirlet G, Dierich A, Depaulis A, Le Meur M, Borrelli E (1995): Parkinsonian-like locomotor impairment in mice lacking dopamine D_2 receptors. Nature 377:424–428.

Banbury Conference on Genetic Background in Mice (1997): Neuron 19:755–759.

Barlow C, Hirotsune S, Paylor R, Liyanage M, Eckhaus M, Collins F, et al. (1996): *Atm*-deficient mice: A paradigm of Ataxia Telangiectasia. Cell 86:159–171.

Behan D, Grigoriadis D, Lovenberg T, Chalmers D, Heinrichs S, Liaw C, De Souza EB (1996): Neurobiology of corticotropin releasing factor (CRF) receptors and CRF-binding protein: implications for the treatment of CNS disorders. Mol Psychi 1:265–277.

Britton KT, Lee G, Dana R, Risch SC, Koob GF (1986): Activating and "anxiogenic-like" effects of corticotropin releasing factor are not inhibited by blockade of the pituitary-adrenal system with dexamethasone. Life Sci 39:1281–1286.

Campbell IL, Gold LH (1996): Transgenic modeling of neuropsychiatric disorders. MolPsychi 1:105–120.

Crawley JN, Belknap JK, Collins A, Crabbe JC, Frankel W, Henderson N, et al. (1997): Behavioral phenotypes of inbred mouse strains: implications and recommendations for molecular studies. Psychopharmacologia 132:107–124.

Crawley JN, Gerfen C, McKay R, Rogawski MA, Sibley DR, Skolnick P. Current Procotols in Neuroscience: John Wiley & Sons, Inc.: New York, 1999.

Crawley JN, Paylor R (1997): A proposed test battery and constellations of specific behavioral paradigms to investigate the behavioral phenotypes of transgenic and
knockout mice. Hormone Beh 31:197–211.

Drago F, Contarino A, Busa L (1999): The expression of neuropeptide-induced exces-
sive grooming behavior in dopamine D_1 and D_2 receptor-deficient mice. Eur J Pharm
365:125–131.

Erickson JC, Clegg KE, Palmiter RD (1996a): Sensitivity to leptin and susceptibility to
seizures of mice lacking neuropeptide Y. Nature 381:415–421.

Erickson JC, Hollopeter G, Palmiter RD (1996b): Attenuation of the obesity syndrome
of *ob/ob* mice by the loss of neuropeptide Y. Science 274:1704–1707.

Gavériaux-Ruff C, Matthes HWD, Peluso J, Kieffer BL (1998): Abolition of morphine-
immunosuppression in mice lacking the μ-opioid receptor gene. PNAS USA
95:6326–6330.

Gehlert DR (1998): Multiple receptors for the pancreatic polypeptide (PP-Fold) family:
physiological implications. P Soc Exp M 28:7–22.

Gerald C, Walker MW, Criscione L, Gustafson EL, Batzl-Hartmann C, Smith KE,
(1996): A receptor subtype involved in neuropeptide-Y-induced food intake. Nature
382:168–171.

Giros B, Jaber M, Jones SR, Wightman RM, Caron MG (1996): Hyperlocomotion and
indifference to cocaine and amphetamine in mice lacking the dopamine transporter.
Nature 379:606–612.

Heinrichs S, Merlo Pich E, Miczek K, Britton K, Koob G (1992): Corticotropin-releasing factor antagonist reduces emotionality in socially defeated rats via direct neu-
rotropic action. Brain Res 581:190–197.

Heinrichs SC, Min H, Tamraz S, Carmouche M, Boehme SA, Vale WA (1997): Anti-
sexual and anxiogenic behavioral consequences of corticotropin-releasing factor
overexpression are centrally mediated. Psychoneuroendocrinology 22:215–224.

Hoebel GB; Wellman PJ. Overview of methodological approaches to the study of in-
gestive behaviors. In Current Protocols in Neuroscience; Crawley JN, Gerfen CR,
McKay R, Rogawski M, Sibley DR, Skolnick P, Eds. John Wiley and Sons, Inc.: New
York, 1998; 8.6A.1–8.6A5.

Holcomb L, Gordon MN, McGowan E, Yu X, Benkovic S, Jantzen P, et al. (1998): Ac-
celerated Alzheimer-type phenotype in transgenic mice carrying both mutant *amyloid precursor protein* and *presenilin 1* transgenes. NatMed 4:97–100.

Hollopeter G, Erickson JC, Seeley RJ, Marsh DJ, Palmiter RD (1998): Response of neuropeptide Y-deficient mice to feeding effectors. Regul Pept 75–76:383–389.

Hökfelt T, Bartfai T, Crawley J (1998): Galanin: Basic Research Discoveries and Therapeutic Implications. Ann NY Acad Sci 863:pp 469.

Inui A (1999): Neuropeptide Y feeding receptors: Are multiple subtypes involved?
Trends Pharmacol Sci 20:43–46.

Irwin S (1968): Comprehensive observational assessment: A systematic, quantitative
procedure for assessing the behavioural and physiologic state of the mouse. Psychopharmacologia 13:222–257.

Jucker M, Ingram DK (1997): Murine models of brain aging and age-related neurodegenerative diseases. Beh Brain Res 85:1–25.

Kieffer BL (1999): Opioids: first lessons from knockout mice. Trends Phar 20:19–26.
Kitanaka N, Sora I, Kinsey S, Zeng Z, Uhl GR (1998): No heroin or morphine 6β-
gluccuronide analgesia in μ-opioid receptor knockout mice. Eur J Pharm
355:R1–R3.

Kitchen I, Slowe SJ, Matthes HWD, Kieffer B (1997): Quantitative autoradiography of μ-, δ-, and κ-opioid receptors in knockout mice lacking the μ-opioid receptor gene. Brain Res 778:73–88.

Kushi A, Sasai H, Koizumi H, Takeda N, Yokoyama M, Nakamura M (1998): Obesity and mild hyperinsulinemia found in neuropeptide Y-Y1 receptor-deficient mice. PNAS USA 95:15659–15664.

Lijam N, Paylor R, McDonald MP, Crawley JN, Deng CX, Herrup K, et al. (1997): Social interaction and sensorimotor gating abnormalities in mice lacking *Dvl1*. Cell 90:895–905.

Livy DJ, Wahlsten D (1997) Retarded formation of the hippocampal commissure in embryos from mouse strains lacking a corpus callosum. Hippocampus 7:2–14.

Loh HH, Liu HC, Cavalli A, Yang W, Chen YF, Wei LN (1998): μ-Opioid receptor knockout mice: effects on ligand-induced analgesia and morphine lethality. Mol Brain Res 54: 321–326.

Lundkvist J, Chai Z, Tehheranian R, Hasanvan H, Bartfai T, Jenck F, et al. (1996): A non-peptidic corticotropin releasing factor receptor antagonist attenuates fever and exhibits anxiolytic-like activity. Eur J Pharm 309:195–200.

Maldonado R, Salardi A, Valverde O, Samad TA, Roques BP, Borrelli E (1997): Absence of opiate rewarding effects in mice lacking dopamine D2 receptors. Nature 388:586–589.

Markel P, Shu P, Ebeling C, Carlson GA, Nagle DL, Smutko JS, et al. (1997): Theoretical and empirical issues for marker-assisted breeding of congenic mouse strains. Nat Genet 17:280–284.

Marsh DJ, Hollopeter G, Kafer KE, Palmiter RD (1998): Role of the Y5 neuropeptide Y receptor in feeding and obesity. Nat Med 4:718–721.

Matthes HWD, Maldonado R, Simonin F, Valverde O, Slowe S, Kitchen I, et al. (1996): Loss of morphine-induced analgesia, reward effect and withdrawal symptoms in mice lacking the μ-opioid-receptor gene. Nature 383:819–823.

Matthes HWD, Smadja C, Valverde O, Vonesch JL, Foutz AS, Boudinot E, et al. (1998): Activity of the δ-opioid receptor is partially reduced, whereas activity of the μ-receptor is maintained in mice lacking the μ-receptor. J Neurosci 18:7285–7295.

Meisel RL, Sachs BD. The physiology of male sexual behavior. In The Physiology of Reproduction, Second Edition; Knobil E, Neill JD, Eds.; Raven Press: New York, 1994; pp 3–5.

Montkowski A, Poettig M, Mederer A, Holsboer F (1997): Behavioural performance in three substrains of mouse strain 129. Brain Res 762:12–18.

Moser VC, Cheek BM, MacPhail RC (1995): A multidisciplinary approach to toxicological screening: III. Neurobehavioral toxicity. J Tox Env H 45:173–210.

Nelson RJ, Demas GE, Huang PL, Fishman MC, Dawson VL, Dawson TM, et al. (1995): Behavioural abnormalities in male mice lacking neuronal nitric oxide synthase. Nature 378:383–386.

Nelson RJ, Young KA (1997): Behavior in mice with targeted disruption of single genes. Neurosci Biobehav Rev 22:453–462.

Ogawa S, Lubahn DB, Korach KS, Pfaff DW (1997): Behavioral effects of estrogen receptor gene disruption in male mice. PNAS USA 94:1476–1481.

Paylor R, Nguyen M, Crawley JN, Patrick J, Beaudet A, Orr-Urtreger A (1998): α7 nicotinic receptor subunits are not necessary for hippocampal-dependent learning or sensorimotor gating: A behavioral characterization of *Acra7*-deficient mice. Learn Mem 5:302–316.

Ramirez VD (1992): Characterization of membrane actions of steroids. NeuroReport 1:35–41.

Rissman EF, Early AH, Taylor JA, Korach KS, Lubahn DB (1997): Estrogen receptors are essential for female sexual receptivity. Endocrinology 138:507–510.

Rogers DC, Fisher EMC, Brown SDM, Peters J, Hunter AJ, Martin JE (1997): Behavioral and functional analysis of mouse phenotype: SHIRPA, a proposed protocol for comprehensive phenotype assessment. Mamm Gen 8:711–713.

Rubenstein M, Phillips TJ, Bunzow JR, Falzone TL, Dziewczapolski G, Zhang G, et al. (1997): Mice lacking dopamine D4 receptors are supersensitive to ethanol, cocaine, and methamphetamine. Cell 90:991–1001.

Sango K, Yamanaka S, Hoffmann A, Okuda Y, Grinberg A, Westphal H, McDonald MP, et al. (1995): Mouse models of Tay-Sachs and Sandhoff diseases differ in neurologic phenotype and ganglioside metabolism. Nat Genet 11:170–176.

Sibley DR, Hollon TR, Gleason TC, Lachowicz JE, Ariano MA, Huang SP, Westphal H, Surmeier DJ, Crawley JN (1998): Generation and characterization of D_5 dopamine receptor knock-out mice. American College of Neuropsychopharmacology 37th Annual Meeting, Abstract 56, page 287.

Silver L. Mouse Genetics: Concepts and Applications. Oxford University Press: New York, 1995.

Skutella T, Criswell H, Moy S, Probst JC, Breese GR, Jirikowski GF, Holsboer F (1994): Corticotropin-releasing hormone (CRH) antisense oligodeoxynucleotide induces anxiolytic effects in rat. NeuroReport 5:2181–2185.

Smith DR, Striplin CD, Geller AM, Mailman RB, Drago J, Lawler CP, Gallagher M (1998a): Behavioural assessment of mice lacking D_{1A} dopamine receptors. Neuroscience 86:135–146.

Smith GW, Aubry JM, Dellu F, Contarino A, Bilezikjian LM, Gold LH, Chen R, Marchuk Y, Hauser C, Bentley CA, Sawchenko PE, Koob GF, Vale W, Lee KF (1998b): Corticotrophin releasing factor receptor 1-deficient mice display decreased anxiety, impaired stress response, and aberrant neuroendocrine development. Neuron 20:1093–1102.

Sokoloff P, Schwartz JC (1995): Novel dopamine receptors half a decade later. Trends Phar 16:270–275.

Sora I, Takahashi N, Funada M, Ujike H, Revay RS, Donovan DM, Miner LL, et al. (1997): Opiate receptor knockout mice define μ receptor roles in endogenous nociceptive responses and morphine-induced analgesia. PNAS USA 94:1544–1549.

Stenzel-Poore MP, Heinrichs SC, Rivest S, Koob GF, Vale WW (1994): Overproduction of corticotropin-releasing factor in transgenic mice: A genetic model of anxiogenic behavior. J Neurosc 14:2579–2584.

Sutton R, Koob G, LeMoal M, Rivier J, Vale W (1982): Corticotropin releasing factor (CRF) produces behavioural activation in rats. Nature 297:331–333.

Timpl P, Spanagel R, Sillaber I, Kresse A, Reul JM, Stalla GK, et al. (1998): Impaired stress response and reduced anxiety in mice lacking a functional corticotropin releasing hormone receptor 1. Nat Genet 19:162–166.

Wehner JM, Silva A (1996): Importance of strain differences in evaluations of learning and memory processes in null mutants. Ment Retard Devel Disabil Res Rev 2:243–248.

Wersinger SR, Sannen K, Villalba C, Lubahn DB, Rissman EF, De Vries GJ (1997): Masculine sexual behavior is disrupted in male and female mice lacking a functional estrogen receptor α gene. Horm Beh 32:176–183.

Xu M, Hu XT, Cooper DC, Moratalia R, Graybiel AM, White FJ, et al. (1994a): Elimination of cocaine-induced hyperactivity and dopamine-mediated neurophysiological effects in D1 receptor mutant mice. Cell 79: 945–955.

Xu M, Moratalia R, Gold LH, Hiroi N, Koob GF, Graybiel AM, et al. (1994b): Dopamine D1 receptor mutant mice are deficient in striatal expression of dynorphin and in dopamine-mediated behavioral responses. Cell 79:729–742.

CHAPTER 3

GENETIC ANALYSIS OF RECEPTOR PHENOTYPES

SCOTT BULTMAN, PATRICIA GREEN, and TERRY MAGNUSON

1. Introduction

Cellular receptors arose during metazoan evolution to coordinate cell-cell interactions. The significance of receptors in regulating cell proliferation and differentiation in organisms ranging from *C. elegans* to humans is underscored by a high degree of phylogenetic conservation and the occurrence of duplication and divergence events that led to a large number of receptors in vertebrates. In fact, the mammalian genome encodes well over 1000 distinct receptor proteins that can be grouped into such structurally and functionally diverse families as receptor tyrosine kinases (RTK), type I and type II serine/threonine kinase receptors, G-protein-coupled receptors, Notch receptors, and steroid receptors.

Transgenic and gene-targeting technologies in the mouse have demonstrated the importance of mammalian receptors in myriad biological processes, including embryonic development, neurotransmission, immunology, and carcinogenesis. In addition, classical genetic methods can be utilized to extend our knowledge of individual receptor genes to genetic pathways. A combination of gene transfer and classical genetic strategies will complement biochemical and molecular approaches towards elucidating receptor function and delineating signal transduction pathways. In this chapter, we emphasize the importance of allelic series and the significance of genetics in phenotype interpretation. We also describe how Mendelian and quantitative genetics can be exploited to map modifier loci and how to generate animals carrying mutations in two or more genes.

2. Gene Targeting and Allelic Series

The vast majority of targeted mutations in receptor genes are nulls in which coding sequence has been deleted and replaced with a neomycin phosphotransferase (*neo*) selectable marker (see Chapter 1 for details). Although null mutations are

Genetic Manipulation of Receptor Expression and Function,
Edited by Domenico Accili.
ISBN 0-471-35057-5   Copyright © 2000 Wiley-Liss, Inc.

essential for elucidating gene function, much can be gleaned by analyzing additional alleles. A case in point is the classical dominant white-spotting (*W*) locus, which encodes the c-Kit RTK. Many spontaneous and agent-induced *W* mutations have been recovered based on their semidominant inheritance and visible coat-color phenotypes (Reith and Bernstein, 1991). Null heterozygotes exhibit white spotting on the head and ventrum due to haploinsufficency, whereas null homozygotes die perinatally. In contrast, hypomorphs are viable as homozygotes but have an entirely white coat, macrocytic anemia, mast cell deficiency, and/or sterility. Analysis of the pleiotropic *W* phenotype demonstrated that c-Kit is necessary for the survival and proliferation of three migratory cell lineages: neural crest-derived melanoblasts, hematopoeitic progenitor cells, and primordial germ cells. Therefore, *W* hypomorphic alleles have provided valuable insight into the role of c-Kit in melanogenesis, hematopoiesis, and gametogenesis and contributed to c-Kit being one of the best-understood RTKs.

Molecular analyses of *W* mutations have also been very informative. Although some mutations are regulatory in nature, many are intragenic lesions that perturb *c-kit* coding exons. Moreover, most of the intragenic mutations are point mutations and provide structure-function relationships about specific amino-acid residues in the transmembrane domain, ATP-binding site, and phosphotransferase domain (Reith and Bernstein, 1991). For example, a nonconservative substitution (L790N) in the phosphotransferase domain of W^{42J} confers dominant-negative activity because the mutant receptor can bind—but not cross-phosphorylate—wild-type receptor. As a result, some of the wild-type receptors are sequestered as nonfunctional heterodimers, and the efficiency of the signal transduction pathway is reduced. Accordingly, W^{42J} heterozygotes have a more severe phenotype than do null heterozygotes, displaying anemia and reduced fertility in addition to extensive white spotting.

Most receptor genes will not be amenable to the straightforward type of dominant, visible coat-color screens used to recover *W* mutants. Instead, most receptor mutations likely will be recessive; dominant effects probably will be uncommon or subtle. In this regard, it is fortunate that many receptor genes have been cloned and much progress has been made in transgenic and gene-targeting technologies. For example, inducible gain-of-function and conditional loss-of-function mutations are now possible utilizing the tet and Cre-*loxP* systems, respectively (see Chapters 2 and 10 for details). Generation of an allelic series via gene targeting is exemplified by recent work on fibroblast growth-factor receptor 1 (*Fgfr1*). *Fgfr1* null mutants die during early embryonic development with gastrulation defects, precluding analysis of gene function at later stages of development (Deng, Wynshaw-Boris et al., 1994; Yamaguchi, Harpal et al., 1994). Mouse *Fgfr1* might be expected to play a role in neural crest or skeletal development given that human *Fgfr1* mutations cause craniofacial and skeletal defects in Pfeiffer syndrome (Muenke, Schell et al., 1994; Muenke and Schell, 1995; Passos-Bueno, Richieri-Costa et al., 1998). To resolve this issue, Partanen et al. (1998) utilized three targeting vectors to generate six novel *Fgfr1* alleles (Table 3.1).

Each *Fgfr1* targeting event was associated with insertion of *neo* into the 7th or 15th intron of the locus. Because *neo* insertions sometimes interfere with splicing and reduce the quantity of wild-type transcripts without any other sequence alteration, they can behave as hypomorphic alleles if the targeted gene

Table 3.1. *Fgfr1* **Allelic Series**

Allele(*)	Lesion	*/* Phenotype	*/null Phenotype
n7	*neo* insertion-intron 7	neonatal lethality	50% embryonic lethal
		anterior homeotic transformations limb defects	50% neonatal lethality anterior homeotic transformations[π] limb defects[π]
		craniofacial abnormalities	craniofacial abnormalities[π]
n15	*neo* insertion-intron 15	neonatal lethality anterior homeotic transformations limb defects craniofacial abnormalities	N.A.
IIIc	stop codon in exon IIIc	embryonic lethal gastrulation defects	N.A.
IIIb	stop codon in exon IIIb	viable and fertile no abnormalities	N.A.
IIIbn	stop codon in exon IIIb plus *neo* insertion-intron 7	neonatal lethality anterior homeotic transformations[π] limb defects craniofacial abnormalities	embryonic lethal
Y766F	Y766F autophosphorylation site	viable and fertile posterior homeotic transformations	viable and fertile posterior homeotic transformations

N.A., not analyzed

[π], increased penetrance and/or severity relative to */*

is sensitive to dosage (Meyers, Lewandoski et al., 1998). In the case of *Fgfr1*, both *neo* insertions caused a three- to five-fold reduction in full-length mRNA and conferred altered homeo box (*Hox*) gene expression, aberrant somitogenesis, anterior transformations and truncations of skeletal elements, limb-patterning, and craniofacial defects, as well as neonatal lethality in the homozygous condition. The *neo* cassettes, which were flanked by *loxP* sites, were subsequently excised by mating *neo* heterozygotes to transgenic mice ubiquitously expressing Cre recombinase. Reversion to wild-type confirmed that the mutant phenotype was due to *neo*. It is possible even to extend these results by utilizing transgenic mice expressing *cre* in a more restricted manner to direct tissue-specific rescue of *Fgfr1* (see Chapter 13 for details).

Each of the three *Fgfr1* targeting vectors contained a point mutation in an exon on the 5′ or 3′ arm of homology. The point mutations were incorporated in some embryonic stem (ES) cell clones but not others depending on where homologous recombination occurred between the targeting vector and the endogenous locus. Two of the targeting vectors were used to assess the relative significance of two *Fgfr1* isoforms, which arise by alternative splicing and have distinct ligand-binding characteristics. A premature stop codon was introduced into each of the alternative exons, IIIb and IIIc. The phenotype of IIIc/IIIc homozygotes recapitulated the null indicating that the IIIc isoform is of critical importance. In contrast, IIIb/IIIb homozgotes could not be distinguished from wild-type indicating that the IIIb isoform is less important. The IIIb isoform is functional, however, because the *neo* hypomorphic allele was associated with a more severe phenotype when the IIIb point mutation was also present. The third targeting vector introduced a missense point mutation into the tyrosine autophosphorylation site (Y766F) that stimulates intracellular PLC-γ and PKC signaling. Heterozygotes and homozygotes were viable and fertile and displayed only posterior homeotic transformations. Important to note, the homeotic transformations observed were opposite in direction to the other alleles and the mutation is dominant, suggesting it is a gain-of-function allele. Thus, the autophosphoylation site and PLC-γ- and PKC-pathway are implicated in mediating receptor down-regulation. Finally, it should be noted that another advantage of having multiple alleles is that compound heterozygotes can be generated in which two distinct mutant alleles (e.g., null/hypomorph) produce novel, often intermediate, phenotypes revealing even more about gene function.

Embryological manipulations can also be utilized to extend what is known about a particular mutant allele. For example, aggregation chimeras consisting of a mixture of wild-type cells expressing a *lacZ* transgene and *Fgfr1* null cells revealed that mutant cells do not properly traverse the primitive streak. Also, these chimeras identified extraembryonic, cephalic, heart, axial and paraxial mesoderm, and endoderm tissues as being particularly dependent on *Fgfr1* function (Ciruna, Schwartz et al. 1997). Aggregation chimeras also address whether a gene product functions cell autonomously.

An alternative strategy for generating allelic series in receptor genes is *N*-ethyl-*N*-nitrosourea (ENU) mutagenesis. ENU is a potent chemical mutagen in the mouse germline, inducing 1 mutation per locus for approximately every 650 gametes screened (Rinchik, 1991). Due to space and financial constraints, however, most in vivo screens have been conducted at institutions with large mouse colonies such as the Oak Ridge National Laboratory, Tennessee, U.S.A., and the MRC Radiobiology Unit in Harwell, England. To circumvent this limitation, ENU mutagenesis can be performed in vitro in ES cells. Hundreds to thousands of mutagenic ES cell colonies can be analyzed by PCR amplifying part or all of a gene of interest and by sequencing the products with high-throughput capillary or chip methodologies. Mutant ES cells subsequently can be used to make chimeras to transmit the mutation through the germline. As a result, mutations created in vitro can have the corresponding phenotype analyzed in vivo. Preliminary evidence suggests that such a scheme is feasible. ENU has been shown to induce mutations at the selectable hypoxanthine guanine phosphoribosyl transferase (*Hprt*) locus in ES cells (Chen Y, Yee D, Woychik R, Magnuson T, unpublished data), and, with an optimized dose, one mutant colony can be recovered per 200 plated cells (Chen Y, Yee D, Woychik R, Magnuson, T, unpublished data). Furthermore ENU-induced *Hprt* mutant ES cells are germline competent (Chen Y, Yee D, Woychik R, Magoon T).

ENU mutagenesis offers several advantages over targeted mutagenesis. First, whereas most targeted mutations have been null alleles, ENU can generate gain-of-function, hypomorph, or antimorph alleles in addition to null alleles. Second, most lesions are point mutations and therefore provide valuable structure-function information. Third, ENU-induced point mutations occur randomly throughout genes in contrast to targeted mutations, which are usually confined to well-characterized motifs or domains. Molecular analysis of ENU-induced and spontaneous mutations have identified functionally important regions that lie outside of conserved domains, which probably would never have been targeted. Worth noting, however, is that, because the screen described above is genotype-driven instead of phenotype-driven, some nucleotide substitutions are expected not to confer a mutant phenotype.

3. Influence of Genetic Background on Phenotype Interpretation

A targeted mutation can be established on an inbred genetic background by breeding chimeras that transmit the mutation at a relatively high frequency to the appropriate 129 substrain (i.e., the same substrain from which the ES cell line was derived). Mutant phenotypes are often more pronounced on inbred genetic backgrounds, and phenotypic analyses can be more straightforward due to less variability. Analyzing the mutation on an inbred 129 background allows the 129 parental strain to serve as a coisogenic, wild-type control. However, it is also wise to introduce targeted mutations onto different genetic backgrounds.

The penetrance and expressivity of mutant phenotypes can be greatly influenced by genetic background. One of the first documented cases investigated more than 60 years ago involves the endothelin receptor B (*Ednrb^s*) mutation, which causes variable piebald spotting on different genetic backgrounds. Although such variability may confound initial phenotypic analyses, it has several important benefits. First, when mouse models of human disease are developed, a mutation can be crossed onto different genetic backgrounds in an attempt to "fine-tune" the phenotype such that the time of onset and pathology more closely resemble a human disease or syndrome (Smithies, 1993). Second, novel phenotypes provide a more complete understanding of gene function. Third, modifier genes responsible for the phenotypic variation can be mapped and possibly cloned. This possibility is important because modifier genes are likely to encode proteins that function in either the same or parallel biochemical pathway as the disrupted gene product. In fact, modifiers have already been mapped for several different mutations, including cystic fibrosis transmembrane conductance regulator homolog (*Cftr*) and transforming growth factor –1 (Han, 1990; Bonyadi, 1997; Rozmahel, 1996).

Another potential approach for identifying interacting loci is to isolate suppressors or enhancers of phenotypes by using ENU mutagenesis. Such screens have been successful in other model organisms such as *C. elegans* and *Drosophila* and have identified novel members of the *dpp*, *sevenless*, *let-23*, and

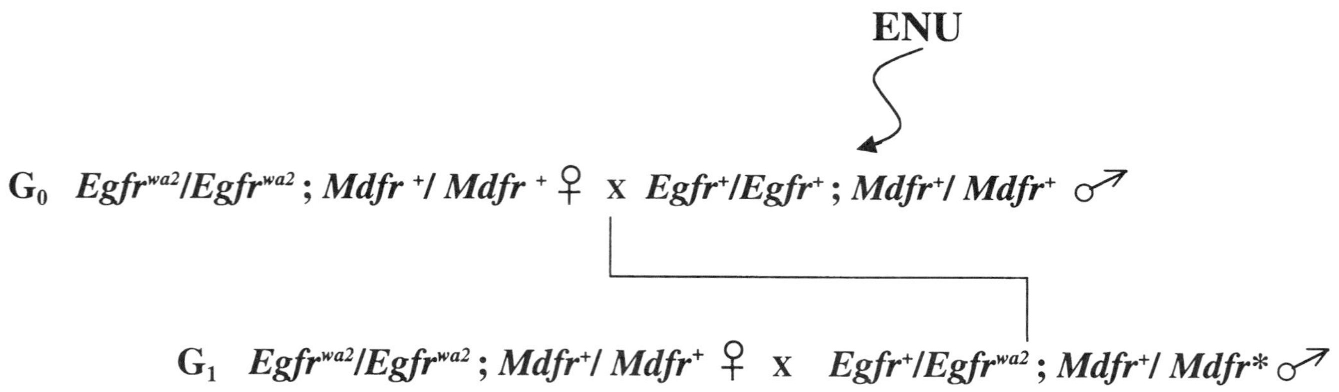

ENU

G$_0$ $Egfr^{wa2}/Egfr^{wa2}$; $Mdfr^+/Mdfr^+$ ♀ x $Egfr^+/Egfr^+$; $Mdfr^+/Mdfr^+$ ♂

G$_1$ $Egfr^{wa2}/Egfr^{wa2}$; $Mdfr^+/Mdfr^+$ ♀ x $Egfr^+/Egfr^{wa2}$; $Mdfr^+/Mdfr^*$ ♂

G$_2$

	$Egfr^{wa2}$; $Mdfr^+$
$Egfr^+$; $Mdfr^+$	Non-waved
$Egfr^{wa2}$; $Mdfr^+$	Waved
$Egfr^+$; $Mdfr^*$	Non-waved
$Egfr^{wa2}$; $Mdfr^*$	Waved

Assay:
Increased fur waviness
Decreased fur waviness
Open eye penetrance
Viability

Figure 3.1. Suppressor/Enhancer screen. Outline of two-generation ENU mutagenesis to isolate suppressors and enhancers of $Egfr^{wa2}$ phenotype. G$_0$ mutagenized males are mated to $Egfr^{wa2}$ homozygotes. G$_1$ males are mated to $Egfr^{wa2}$ females. Some G$_1$ males will carry mutations in modifier gene loci ($Mdfr^*$). In the G$_2$ generation, mutations that enhance or suppress $Egfr^{wa2}$ phenotype can be isolated due to changes in the waved phenotype, including fur waviness, viability, and presence of open eyes.

other receptor pathways (Han, Aroian et al., 1990; Rogge, Karlovich et al., 1991; Simon, Bowtell et al., 1991; Chen, Riese et al., 1998). However, application of this technique to the mouse requires tight control of genetic background; otherwise phenotypic variability could be due to changes in background instead of second-site suppressor or enhancer mutations.

There is evidence that suppressor/enhancer screens are feasible in the mouse. For example, it should be possible to isolate novel members of murine *Egfr* signaling pathways by exploiting the classical waved-2 (*Egfr^{wa2}*) mutant stock. This hypomorph of *Egfr* has a point mutation in the intracellular tyrosine kinase domain (Luetteke, Phillips et al., 1994; Fowler, Walker et al., 1995). Homozygotes are viable and fertile but have waved fur and curly vibrissae. Sporadically, they exhibit open eyes at birth. Important to note is that this phenotype is enhanced such that 20% of *Egfr^{wa2}*/*Egfr^{wa2}* pups exhibit open eyes if they are also heterozygous for a mutation in son of sevenless homolog 1 (*Sos1*) (Wang, Hammond et al., 1997). In addition, there is 50% subviability. *Sos1* encodes a guanine nucleotide exchange factor downstream of *Egfr* (Webb, Jenkins et al., 1993). These data indicate that the waved-2 phenotype is dosage-sensitive and amenable to a suppressor/enhancer screen. Thus, new mutations could be induced that interact with the waved mutation by screening for detectable changes in waviness, open eyes, or viability. Such a screen would require mating ENU mutagenized males to *Egfr^{wa2}*/*Egfr^{wa2}* females (Fig. 3.1). The offspring are mated to *Egfr^{wa2}*/*Egfr^{wa2}* and the pups from this cross are screened for new phenotypes. Whereas this technique is generally feasible only for homozygous viable mutations that exhibit dosage sensitivity, it is a powerful tool for isolating genetic pathway members.

4. Mapping Modifier Loci

Improved genotyping methods and a dense genetic map have made it possible to map modifier genes (for example, Dietrich, Lander et al., 1993; MacPhee, Chepenik et al., 1995; Rozmahel, Wilschanski et al., 1996; Bonyadi, Rusholme et al., 1997). To map modifiers, two strains are exploited that display different degrees of phenotypic severity. Although crosses between the two strains may result in a continuous spectrum of phenotypes, individuals at the phenotypic extremes, scored as "phenotype A," characteristic of Strain A, or "phenotype B," characteristic of Strain B, are used. Modifiers (Mdfr) responsible for the phenotypic extremes can be genetically mapped by analyzing the co-segregation of strain-specific dinucleotide repeats called microsatellite markers (also known as simple sequence length polymorphisms, or SSLPs) with phenotypic extremes in mapping crosses.

Mapping crosses may be of two types: an F_1 intercross or an N_1 backcross (Fig. 3.2). Strain A heterozygotes (often 129/Sv for targeted mutations) are mated to Strain B. The heterozygous progeny are either sib mated (intercross, Fig. 3.2B) or bred to Strain A heterozygotes (backcross, Fig. 3.2A). For the intercross, SSLP segregation at any marker locus among the offspring is 1 Strain A/Strain A: 2 Strain A/Strain B: 1 Strain B /Strain B, whereas segregation is 1 Strain A/Strain B: 1 Strain A/Strain A for a backcross. An intercross has the advantage of providing more meiotic information per individual analyzed (and

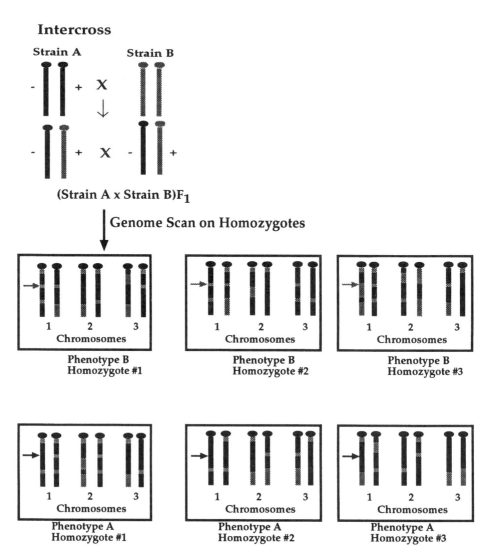

Figure 3.2. Mapping modifier loci. (A) Backcross: Strain A (black chromosomes, often 129/Sv) heterozygous mice are outcrossed to Strain B (light chromosomes) mice to produce (Strain A x Strain B)F$_1$ mice, which contain one set of Strain A chromosomes and one set of Strain B chromosomes. F$_1$ mice, which are heterozygous for the targeted mutation, are backcrossed to Strain A heterozygotes to produce wild-type ($+/+$), heterozygous ($+/-$) and homozygous ($-/-$) progeny. These animals will have one set of Strain A chromosomes (black) and one set of chromosomes that are a mix of Strain A (black) and Strain B (light) due to meiotic recombination in the F$_1$ germline. If homozygous animals showing a phenotype A versus phenotype B are detected in this generation, a Strain B allele at one or more modifier loci acting in a dominant manner is responsible for the phenotype B condition. A genome scan is performed on the phenotype B homozygotes with microsatellite markers polymorphic between Strain A and Strain B. The goal is to identify regions in common of Strain B origin in the phenotype B homozygotes (light arrows for chromosome 1). These regions likely contain the Strain B allele necessary for phenotype B. To confirm these results, a genome scan on the phenotype A homozygotes is performed. At least a portion of the regions of Strain B required for phenotype B should be of Strain A origin (dark arrows for chromosome 1).

Backcross

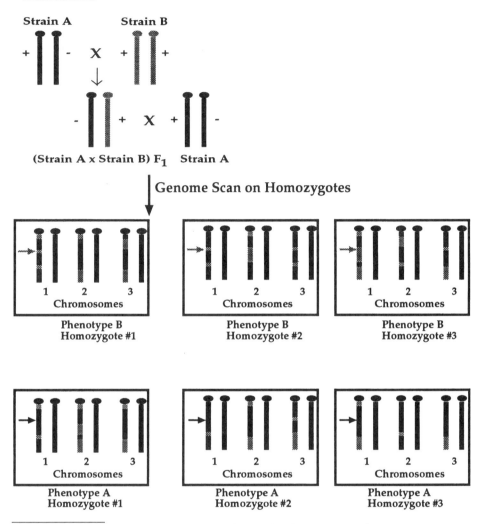

Genome Scan on Homozygotes

(B) Intercross: Strain A (black chromosomes, often 129/Sv) heterozygous mice are out-crossed to Strain B (light chromosomes) mice to produce (Strain A x Strain B)F_1 mice, which contain one set of Strain A chromosomes and one set of Strain B chromosomes. F_1 heterozygotes are crossed *inter se* to produce wild-type (+/+), heterozygous (+/−) and homozygous (−/−) animals in the next generation. These animals will have two sets of chromosomes that are a mix of Strain A (black) and Strain B (light) due to mei-otic recombination in the F_1 germline. If homozygous animals showing a phenotype A versus phenotype B are detected in this generation, a Strain B allele at one or more mod-ifier loci is acting either in a dominant or recessive manner. A genome scan is performed on phenotype A versus phenotype B homozygotes with microsatellite markers poly-morphic between Strain A and Strain B. The goal is to identify regions in common with Strain B origin in phenotype B homozygotes (light arrows) that are not present in Strain A homozygotes (dark arrows) as described in A. These regions likely contain Strain B alleles that are responsible for rescuing the phenotype. In this example, a recessive Strain B modifier is depicted and the region highlighted by the light arrow is of Strain B origin for both chromosomes.

hence fewer mice are needed) because both parents contribute informative meioses. On the one hand, the statistical analysis is more difficult because there are more genotypic classes of progeny. On the other hand, a backcross requires more individuals per mapping experiment because only one parent generates informative meioses, but the statistical analysis is simplified because there are fewer genotypic classes. Also, a backcross identifies only dominant or co-dominant Strain B modifiers.

Genome scans allow linkage analysis with standard mapping programs such as Mapmaker QTL (Lander, Green et al., 1987). Statistically significant segregation of Strain B alleles of microsatellite markers with the "B" phenotypic extreme indicates linkage. Ordinarily, a p-value of <0.05 is sufficient to reject the null hypothesis (that is, no linkage between the marker locus and putative modifiers exists) and infer linkage. However, the large number of tests genome wide (50 or greater for 30-cM resolution) means that a p value of 0.05 may occur by chance two to three times in a genome scan. Thus, Lander and Kruglyak (1995) have suggested using more stringent p values to account for this fact. Suggestive linkage is given by $p < 3.4 \times 10^{-3}$ for a backcross and $p < 1.6 \times 10^{-3}$ for an intercross.

The number of mice collected for genome scans depends upon the number of modifiers mediating the effect. To estimate this number, a backcross can be used. For example, if 1-dominant unlinked modifier causes phenotype B, it is simple to predict the percent of F_1 offspring [Strain A $+/-$ X Strain B $+/-$]$+/-$ X Strain A $+/-$ that will exhibit phenotype B. One-fourth of the offspring will be $-/-$; and 1/2 will be $Mdfr^B/+$, and thus $\sim1/8$ (12.5%) of offspring will display phenotype B. For 2-dominant modifiers, $\sim6.25\%$ of offspring with phenotype B are expected. Theoretically, for 1 or 2 modifiers with relatively large effects, 40 informative meioses with 20-cM intervals should be sufficient for linkage analysis (Lander and Schork, 1994). In practice, for 1 or 2 modifiers with relatively large phenotypic consequences, a sample size as small as 50 individuals from an intercross and 30–35 cM marker intervals has provided significant linkage evidence (Bonyadi, Rusholme et al., 1997).

As stated, the piebald ($Ednrb^s/Ednrb^s$) phenotype was determined to vary with genetic background more than 60 years ago (Dunn, 1937; Reed, 1937). Although at that time it was impossible to map such modifiers, selective breeding established a highly variable phenotype due to a set of modifiers, called the k-complex. Using microsatellite mapping techniques, Pavan and coworkers (1995) could map the responsible modifiers. Piebald mice from a C3H background, which show minimal spotting, were crossed to Mayer background mice with extensive spotting. The F_1 progeny resembled the C3H background with minimal spotting and were backcrossed to Mayer piebald mice. N_2 piebald offspring were categorized according to the degree of spotting. Initially, a 30-cM linkage analysis implicated six genomic regions in greater spotting. Segregation of the initial markers and tightly linked markers in 90 additional offspring from the phenotypic extremes (highly spotted or not spotted) demonstrated that 4 of the 6 regions, on chromosomes 2, 5, 8, and 10, maintained p values of <0.001 when this larger group was included. Three of the regions implicated in the linkage analysis suggest interesting candidate loci. The marker with closest linkage on chromosome 5 maps within 1 cM of three dominant spotting mutations, $W(c\text{-}kit)$, Patch (platelet derived growth factor receptor) and Rump-white. The chromosome 8 modifier lies near extension (e), which codes the melanocyte-stimulating hormone receptor and affects the ratio of eumelanin to phaeomelanin in melantocytes (Robbins, Nadeau et al., 1993). Consistent with a modifier role, e mutations suppress the spotting in piebald mice (Lamoreux and Russell, 1979). Finally, the modifier on chromosome 10 lies near the Steel locus, which encodes the mast cell-growth factor that is the ligand for c-Kit.

In contrast to the piebald example, genome scan results are not always straightforward. The targeted $Egfr$ null phenotype also varies with genetic background, and homozygotes display peri-implantation lethality on a CF-1 background, midgestational lethality on a 129/Sv background, and perinatal lethality on a CD-1 or C57BL/6J background (Sibilia and Wagner, 1995; Threadgill, Dlugosz et al., 1995). In a backcross of [129/Sv X CD-1] N_1 heterozygotes to 129/Sv heterozygotes, $\sim5\%$ of offspring survive beyond midgestation, suggesting that 2–3 dominant or semidominant modifiers produce this effect. A 30-cM genome scan of homozygotes, which are either phenotypically normal or dying at midgestation, demonstrate that the genetic basis of the phenotypic differences is quite complex. Five genomic regions, located on chromosomes 5, 7, 15, and 17 show pointwise statistical significance ($p < 0.05$) with survival at midgestation. However, increasing the sample size for these markers did not increase the pointwise statistical significance to that of suggestive linkage for genome-wide studies. Because there is no overlap in phenotypes on different backgrounds, nor intermediate phenotypes observed, it is somewhat surprising that phenotypic variation is not mediated by one or two strong modifiers. Thus, the normal phenotype at midgestation apparently is mediated by a combination of many modifiers and not by a few strong modifiers.

For us to gain insight into signal transduction pathways underlying biological processes, modifier genes must be cloned and characterized at the molecular level. Although cloning single-gene or major modifiers recently has been met with success in the case of $Mom1$ (a phosholipase that modifies the number of intestinal tumors conferred by Apc^{Min1}), mahogany (a transmembrane receptor related to attractin that modifies the coat-color and obesity phenotype of agouti-locus mutations) (Gunn, Miller et al., 1999; Nagle, McGrail et al., 1999), and $Mvwf$ (a glycotransferase that modifies murine von Willebrand factor) (Mohlke, Purkayastha et al., 1999), it will be much more difficult to clone minor modifiers. First, it is difficult to map minor modifiers as exemplified by the aforementioned $Egfr$ screen. Second, although intriguing candidate genes can be readily selected, it is difficult to demonstrate unequivocally which, if any, are involved. In many cases, only a subtle change in expression level or coding sequence will distinguish sensitive versus resistant strains. More important, performing a knock-out or knock-in experiment on any given gene might be obfuscated by the presence of the other modifier genes. For example, a mutant phenotype might be modified by 5–10 genes where any one, two, or three are dispensable if the other genes are represented by the appropriate alleles. Validation of modifier genes will also be complicated by the fact that some may have other functions and might cause lethal phenotypes if perturbed. Consequently, it seems likely that validation of candidate genes will be the crux of mapping and cloning modifier genes.

5. Redundancy and Analysis of Multiple Mutations

A recurring theme in gene targeting has been that mutations often result in surprisingly mild phenotypes or no phenotype at all. In many cases, the targeted gene arose through duplication and subsequent divergence of an ancestral gene, suggesting that other, presumably related, gene(s) can compensate in its absence. Multiple gene knockout and knock-in experiments have supported this hypothesis, which was proposed by Muller in 1935 to explain a viable two-gene deficiency in *Drosophila*. Multiple gene knockouts have also helped establish or reshape conceptual frameworks for how some of the corresponding proteins are thought to function. For example, breeding mice with mutations in different retinoic acid receptor genes (*Rara*, *Rarb*, and *Rarg* as well as *Rxra*, *Rxrb*, and *Rxrg*) demonstrated that RXR-α/RAR heterodimers mediate retinoid signaling in vivo (Kastner, Grondona et al., 1994). Similarly, comparing the relative growth deficiency of single and double mutants enabled predictions to be made regarding insulin-like growth factor (IGF) receptor-ligand interactions (Baker et al., 1993). From embryonic day (E) 11.0–12.5, IGF-II signals *via* the type 1 receptor (IGF1R). From E13.5 onward, however, both IGF-I and IGF-II signal through IGF1R, and IGF-II also signals through a second receptor subsequently shown to be the insulin receptor (Baker et al., 1993; Louvi et al., 1997; Morrione et al., 1997).

Because different targeted mutations are often maintained on different, sometimes mixed, genetic backgrounds, the more severe phenotype of a "double mutant" might be due to changes in genetic background rather than the mutations in question. Therefore, each genotypic class of progeny should be analyzed to control for differences in genetic background. In a double-heterozygous intercross, for example, 2 out of 16 progeny will be homozygous for one gene while being wild-type for the other $(+/+, -/-$ or $-/-, +/+)$. These mice, along with other predicted classes of progeny $(+/-, +/-; +/-, -/-; -/-, +/-; +/+, +/+)$, should be compared with double homozygotes $(-/-, -/-)$ to evaluate the relative contribution of each mutation toward the mutant phenotype. Analysis of the different classes might also uncover dosage-dependent quantitative effects such that the mutant phenotype becomes progressively more severe as wild-type alleles are replaced by mutant alleles. Such intergenic interactions can be additive or synergistic in nature and suggest that the gene products interact. Data of this type can be valuable for designing biochemical experiments such as two-hybrid assays or co-immunoprecipitation to determine whether the interaction is direct. However, a potential caveat to this approach is that two mutations may compromise the general health of a common subset of cells by unrelated mechanisms and that these cells in double mutants exhibit a more severe phenotype or even die.

6. Future Prospects

Although the ability to exploit different genetic backgrounds to manipulate phenotypes of mutant mice has existed for more than half a century, the advent of microsatellite markers and a dense genetic map have made it possible to map modifier loci. Utilizing the complete sequence of the mouse genome should enable these genes to be cloned. Analysis of modifier genes coupled with classical and targeted mutagenesis to create allelic series and strains with multiple mutations will undoubtedly reveal genetic pathways and molecular mechanisms underlying receptor function and signal transduction pathways.

References

Baker J, Liu JP, Robertson EJ, Efstratiadis A (1993): Role of insulin-like growth factors in embryonic and postnatal growth. Cell 75:73–82.

Bonyadi M, Rusholme SA et al. (1997): Mapping of a major genetic modifier of embryonic lethality in TGF beta 1 knockout mice. Nat Genet 15:207–11.

Chen Y, Riese MJ et al. (1998): A genetic screen for modifiers of *Drosophila* decapentaplegic signaling identifies mutations in punt, Mothers against dpp and the BMP-7 homologue, 60A. Development 125:1759–68.

Ciruna BG, Schwartz L et al. (1997): Chimeric analysis of fibroblast growth factor receptor-1 (Fgfr1) function: a role for FGFR1 in morphogenetic movement through the primitive streak. Development 124:2829–41.

Deng CX, Wynshaw-Boris A et al. (1994): Murine FGFR-1 is required for early postimplantation growth and axial organization. Gene Dev 8:3045–57.

Dietrich WF, Lander ES et al. (1993): Genetic identification of Mom-1, a major modifier locus affecting Min-induced intestinal neoplasia in the mouse. Cell 75:631–9.

Dobrovolsky VN, Casciano DA et al. (1996): Development of a novel mouse tk+/− embryonic stem cell line for use in mutagenicity studies. Env Mol Mut 28:483–9.

Dunn LC (1937): Genetic Analysis of variegated spotting in the house mouse. Genetics 22:43–64.

Fowler KJ, Walker F et al. (1995): A mutation in the epidermal growth factor receptor in waved-2 mice has a profound effect on receptor biochemistry that results in impaired lactation. PNAS US 92:1465–9.

Gunn TM, Miller KA et al. (1999): The mouse mahogany locus encodes a transmembrane form of human attractin [In Process Citation]. Nature 398:152–6.

Han M, Aroian RV et al. (1990): The let-60 locus controls the switch between vulval and nonvulval cell fates in *Caenorhabditis elegans*. Genetics 126:899–913.

Kastner P, Grondona JM et al. (1994): Genetic analysis of RXR alpha developmental function: convergence of RXR and RAR signaling pathways in heart and eye morphogenesis. Cell 78:987–1003.

Lamoreux ML, Russell ES (1979): Developmental interaction in the pigmentary system of mice. I. Interactions between effects of genes on color of pigment and on distribution of pigmentation in the coat of the house mouse (*Mus musculus*). J Heredity 70:31–6.

Lander E, Kruglyak L (1995): Genetic dissection of complex traits: guidelines for interpreting and reporting linkage results [see comments]. Nat Genet 11:241–7.

Lander ES, Green P et al. (1987): MAPMAKER: an interactive computer package for constructing primary genetic linkage maps of experimental and natural populations. Genomics 1:174–81.

Lander ES, Schork NJ (1994): Genetic dissection of complex traits [published erratum appears in Science 1994 Oct 21; 266(5184):353]. Science 265:2037–48.

Luetteke NC, Phillips HK et al. (1994): The mouse waved-2 phenotype results from a point mutation in the EGF receptor tyrosine kinase. Gene Dev 8:399–413.

MacPhee M, Chepenik KP et al. (1995): The secretory phospholipase A2 gene is a candidate for the Mom1 locus, a major modifier of ApcMin-induced intestinal neoplasia. Cell 81:957–66.

Meyers EN, Lewandoski M et al. (1998): An Fgf8 mutant allelic series generated by Cre- and Flp-mediated recombination. Nat Genet 18:135–141.

Mohlke KL, Purkayastha AA et al. (1999): Mvwf, a dominant modifier of murine von Willebrand factor, results from altered lineage-specific expression of a glycosyltransferase. Cell 96:111–20.

Muenke M, Schell U (1995): Fibroblast-growth-factor receptor mutations in human skeletal disorders. Trends Gen 11:308–13.

Muenke M, Schell U et al. (1994): A common mutation in the fibroblast growth factor receptor 1 gene in Pfeiffer syndrome. Nat Genet 8:269–74.

Nagle DL, McGrail SH et al. (1999):. The mahogany protein is a receptor involved n suppression of obesity [In Process Citation]. Nature 398:148–52.

Partanen J, Schwartz L et al. (1998): Opposite phenotypes of hypomorphic and Y766 phosphorylation site mutations reveal a function for Fgfr1 in anteroposterior patterning of mouse embryos. Gene Dev 12:2332–44.

Passos-Bueno MR, Richieri-Costa A et al. (1998): Presence of the Apert canonical S252W FGFR2 mutation in a patient without severe syndactyly. J Med Genet 35:677–9.

Pavan WJ, Mac S et al. (1995): Quantitative trait loci that modify the severity of spotting in piebald mice. Genome Res 5:29–41.

Reed SC (1937): The nheritance and expression of fused, a nuw mutation in the house mouse. Genetics 22:1–13.

Reith AD, Bernstein A (1991): Molecular basis of mouse developmental mutants. Gene Dev 5:1115–23.

Rinchik EM (1991): Chemical mutagenesis and fine-structure functional analysis of the mouse genome. Trend Genet 7:15–21.

Robbins LS, Nadeau JH et al. (1993): Pigmentation phenotypes of variant extension locus alleles result from point mutations that alter MSH receptor function. Cell 72:827–34.

Rogge RD, Karlovich CA et al. (1991): Genetic dissection of a neurodevelopmental pathway: son of sevenless functions downstream of the sevenless and EGF receptor tyrosine kinases. Cell 64:39–48.

Rozmahel R, Wilschanski M et al. (1996): Modulation of disease severity in cystic fibrosis transmembrane conductance regulator deficient mice by a secondary genetic factor [published erratum appears in Nat Genet 1996 May; 13(1):129]. Nat Genet 12:280–7.

Sibilia M, Wagner EF (1995): Strain-dependent epithelial defects in mice lacking the EGF receptor [published erratum appears in Science 1995 Aug 18; 269(5226):909]. Science 269:234–8.

Simon MA, Bowtell DD et al. (1991): Ras1 and a putative guanine nucleotide exchange factor perform crucial steps in signaling by the sevenless protein tyrosine kinase. Cell 67:701–16.

Smithies O (1993): Animal models of human genetic diseases. Trends Gen 9:112–6.

Threadgill DW, Dlugosz AA et al. (1995): Targeted disruption of mouse EGF receptor: effect of genetic background on mutant phenotype. Science 269:230–4.

Wang DZ, Hammond VE et al. (1997): Mutation in Sos1 dominantly enhances a weak allele of the EGFR, demonstrating a requirement for Sos1 in EGFR signaling and development. Gene Dev 11: 309–20.

Webb GC, Jenkins NA et al. (1993): Mammalian homologues of the *Drosophila* Son of sevenless gene map to murine chromosomes 17 and 12 and to human chromosomes 2 and 14, respectively. Genomics 18:14–9.

Yamaguchi TP, Harpal K et al. (1994): fgfr-1 is required for embryonic growth and mesodermal patterning during mouse gastrulation. Genes Dev 8: 3032–44.

CHAPTER 4

IMMORTALIZATION OF CELL LINES FROM TRANSGENIC AND KNOCKOUT MICE BY VIRAL TRANSFORMATION

JANICE YANG CHOU

I. INTRODUCTION

A major goal of molecular cell biology is to study normal gene regulation. Studies in vivo are complicated by our inability to control and manipulate the environment of tissues having specialized functions. As a result, technologies have been developed that allow scientists to generate a variety of in vitro cell models. Immortalized cell lines have been established from tumors (Gey et al., 1952; Pattillo and Gey, 1968; Aden et al., 1979) as well as by transformation of normal cells with chemical carcinogens (Stampfer and Bartley, 1985), oncogenes (Yoakum et al., 1985; Amsterdam et al., 1988; He et al., 1990; Yankaskas et al., 1993), tumor viruses (Vitry et al., 1974; Chou 1978a, 1978b; Isom et al., 1980; Rhim et al., 1985), and by inactivation of tumor-suppressing genes (Gao et al., 1996; Kiyono et al., 1998). These cell lines retain a spectrum of tissue-specific functions and have greatly advanced the study of gene regulation and cellular differentiation. However, results obtained from studies with these transformed cell lines, which possess only the transformed phenotype, may not simulate in vivo gene regulation.

To provide cell models that can propagate in culture, express specialized tissue functions, and simulate in vivo conditions, our laboratory has explored the use of a DNA tumor virus, SV40, to immortalize mammalian cells. The SV40 virus transforms cells of many mammalian species. Whether any given cell type becomes transformed or supports a lytic infection depends on the genetic properties of the host cell (Tooze, 1981). Monkey and human cells are permissive and semipermissive hosts, respectively, for the SV40 virus. Most other mammalian species, including the rodents, are nonpermissive for SV40. On the

Genetic Manipulation of Receptor Expression and Function,
Edited by Domenico Accili.
ISBN 0-471-35057-5   Copyright © 2000 Wiley-Liss, Inc.

one hand, permissive cells support full expression of the viral genome and the release of progeny viral particles, and these cells are killed following SV40 infection. Nonpermissive cells, on the other hand, do not support viral replication, they express some viral early genes, and they can be transformed by SV40. Semipermissive cells support limited lytic infection, release low levels of progeny virions, and can also become transformed by SV40.

Viral transformation generally leads to the dysregulation of the tissue-specific functions of the host cell, and numerous attempts have been made to generate viral-transformed cells that retain differentiated functions. It is generally accepted that a defined culture condition that best supports normal cellular function is a prerequisite for establishing a differentiated, immortalized cell-culture system. For example, hepatocyte lines expressing liver-specific functions have been established by transforming primary rat hepatocytes with a wild-type SV40 virus in a serum-free, hormone-supplemented medium in the presence of an extracellular matrix (Georgoff et al., 1984). Such culture conditions are also essential for continued maintenance of the differentiated phenotype of transformed cells in culture (Fong et al., 1981; Reid and Jefferson, 1984).

When the culture conditions for maintenance of the differentiated phenotype of the host cell are not clearly established, SV40 wild-type virus-immortalized cells express few tissue-specific functions (Vitry et al., 1974). To circumvent this problem and to develop a nonstringent culture condition for maintaining viral-transformed cells in a differentiated state, we developed temperature-sensitive cell models that are reversible for transformation. These cell lines were established by conditionally transforming normal differentiated cells in culture with a *tsA* mutant of SV40 that possesses a large tumor (T) antigen that is functional at one temperature but not another. The T-antigen, which is encoded by the *A* gene of SV40, is required for the initiation and maintenance of transformation (Martin and Chou, 1975; Martin, 1981). We predicted that by manipulating the growth temperature of cells transformed by a SV40 *tsA* mutant, we could reversibly switch between normal and transformed cellular phenotypes. This switch would allow a closer scrutiny of the process of transformation and cell differentiation. To test this hypothesis, we established SV40 *tsA* mutant-transformed human placental cell lines and showed that these transformed cells were temperature-sensitive for differentiation (Chou, 1978a, 1978b). The cells produce low levels of placental proteins, such as human chorionic gonadotropin, pregnancy-specific glycoprotein, and alkaline phosphatase, at the permissive temperature (transformed phenotype). At the nonpermissive temperature, however, these cells assume a normal differentiated phenotype that is characterized by the production of the above-mentioned placental proteins at greatly increased levels. Our results, for the first time, established that SV40 *tsA* mutant virus-transformed cells exhibiting a differentiated phenotype provide a model system that closely simulates normal differentiation in vivo. Most important is that, because a single cell line serves as the source for both transformed and normal states, it is possible, for the first time, to have internally controlled experiments.

Temperature-sensitive cell lines, reversible for transformation, have since been established from fetal (Chou and Schlegel-Haueter, 1981) and adult (Chou, 1983) rat hepatocytes, Schwann cells (Chen et al., 1987), granulosa cells (Zilberstein et al., 1989), and decidual cells (Sirvastava et al., 1995) of rats, mouse macrophages (Takayama et al., 1986), wild-type (Zaret et al., 1988) and albino mutant (Chou et al., 1991) mouse hepatocytes, rabbit endometrial cells (Li et al., 1989), and human epidermal cells (Banks-Schlegel and Howley, 1983). In general, these cell lines are undifferentiated at a permissive temperature and differentiated at a nonpermissive temperature. For example, at a nonpermissive temperature, human epidermal cell line immortalized by the SV40 *tsA*209 mutant exhibits virtually the same degree of terminal differentiation as normal differentiated keratinocytes but becomes de-differentiated at the permissive temperature (Banks-Schlegel and Howley, 1983). Another example is the SV40 *tsA*255 mutant-transformed mouse hepatocyte line, which actively expresses the albumin gene, but only at a nonpermissive temperature (Zaret et al., 1988). This approach has been successfully adapted to generate temperature-sensitive cell lines from a variety of tissues by using transgenic mice expressing a temperature-sensitive T-antigen of the SV40 virus (Whitehead et al., 1993; Cairns et al., 1994; Cluzeaud et al., 1996).

The development of transgenic and knockout technology to manipulate the mouse genome has revolutionized the ability of researchers to generate new animal models. Transgenic technologies have allowed scientists to ectopically express or generate germ-line mutations in virtually any gene in the mouse genome, and knockout technologies have facilitated the development of a variety of mouse models of human diseases. It is important to note that mouse models offer a resource for testing potential treatment strategies before they are used in clinical trials. Moreover, these mouse models provide valuable resources for establishing cell-culture systems harboring the specific mutations. In this report, I discuss the use of *tsA* mutants of SV40 to establish immortalized hepatocyte lines that express liver-specific functions and are reversible for transformation. The techniques are simple and straightforward and should be easily adapted to establish cell lines from various tissues or organs of transgenic and knockout mice.

2. THE SV40 *tsA* MUTANT VIRUSES

Stocks of *tsA* mutants of SV40 are prepared by infecting CV-1 monkey kidney monolayer cells at a low multiplicity of infection (MOI) [~0.01 plaque-forming unit (PFU)/cell] at 34°C (permissive temperature). The cells are allowed to proceed to lysis (Chou and Martin, 1974; Chou, 1985) in α-modified minimal essential medium (αMEM) supplemented with 4% fetal bovine serum (FBS). The infected cultures are frozen, then thawed, and centrifuged to remove cell debris. The lysates, which contain ~10^7 to 10^8 PFU/ml are stored at −20°C. Under the experimental conditions described, it takes approximately two to three weeks for SV40 *tsA* mutant-infected cultures to proceed to lysis. The media of infected cultures are renewed every 3 to 4 days during the first 7 to 10 days after infection. Infected cultures should be allowed to proceed to lysis without additional medium change once cytopathic effects are apparent.

The titer of a SV40 *tsA* mutant is determined by plaque assays (Chou and Martin, 1974). CV-1 monolayers are grown to confluence at 37°C in (MEM containing 4% FBS. The media are removed, and 0.5 ml of a serially diluted virus stock is added per 25-cm^2 flask. Adsorption is allowed to proceed at 34°C for 2 to 3 h with gentle mixing every 15 to 20 min. Then, the viral fluid is aspirated, and 1.2% melted Nobel agar in modified Eagle's medium buffered with tricine (Chou and Martin, 1974) and supplemented with 5% FBS is added and allowed to gel. Fresh medium in agar is added seven days after infection. Neutral red (0.04%) in 1.2% Noble agar is added on day 13 after infection, and plaques are counted on day 14.

3. GENERATION OF G6PASE-DEFICIENT MICE

The current focus in our laboratory is to study the molecular genetics of glycogen storage disease type 1 (GSD-1), also known as von Gierke disease (Chen and Burchell, 1995). It is a group of autosomal recessive disorders characterized by hypoglycemia, hepatomegaly, kidney enlargement, growth retardation, lactic acidemia, hyperlipidemia, and hyperuricemia. GSD-1 is caused by a deficiency in the activity of the endoplasmic reticulum-bound glucose-6-phosphatase (G6Pase), a key enzyme in glucose homeostasis. Structurally, the active site of G6Pase faces the lumen of the endoplasmic reticulum (Pan et al., 1998) and it has been proposed that hydrolysis of glucose-6-phosphate in vivo requires the participation of several integral membrane proteins, including G6Pase and a glucose-6-phosphate transporter (G6PT) (Arion et al., 1980). Accordingly, it was suggested that defects in G6Pase and G6PT correspond to GSD-1a and GSD-1b, respectively (Chen and Burchell, 1995). Recent molecular genetic studies have demonstrated that GSD-1a is caused by mutations in the *G6Pase* gene that abolish or greatly reduce the enzymatic activity (Lei et al., 1993) and GSD-1b is caused by mutations in the *G6PT* gene that abolish glucose-6-phosphate transport activity (Hiraiwa et al., 1999). GSD-1a is the most prevalent form, representing more than 80% of GSD-1 cases. To study the biology and pathophysiology of GSD-1a and to develop novel therapeutic approaches for this disorder, we have generated a G6Pase-deficient (G6Pase$^{-/-}$) mouse by targeted-disruption of the mouse *G6Pase* gene and demonstrated that the phenotype of the G6Pase$^{-/-}$ mice mimics that of GSD-1a patients (Lei et al., 1996). Studies have shown that liver transplants can correct hypoglycemia and other biochemical abnormalities associated with GSD-1a. This finding is not unexpected because G6Pase is expressed primarily in the gluconeogenic organs, such as the liver and the kidney (Nordlie and Sukalski, 1985). Therefore, the G6Pase$^{-/-}$ mice offer a valuable resource from which immortalized hepatocyte lines could be established for the study of GSD-1a and the interrelationship between G6Pase and G6PT.

4. GENE-TARGETING

The *G6Pase* gene contains five exons (Fig. 4.1), spanning ∼12 kb and encoding a polypeptide of 357 amino acid residues (Lei et al., 1993). The G6Pase$^{-/-}$ mice were generated by replacing exon 3 and the flanking introns of the murine

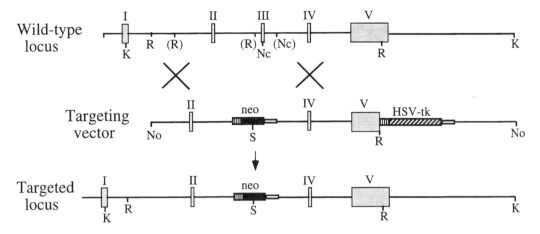

Figure 4.1. Targeted-disruption of the murine *G6Pase* gene. The wild-type locus is presented along with the targeting vector and the anticipated outcome of the recombination (targeted locus). Introns are denoted as lines, exons as boxes. The PGK-1 promoters driving the *neo* and *HSV-tk* genes are denoted as hatched boxes. Gene targeting results in the replacement of exon 3 and the associated introns with a neo cassette (Lei et al., 1996). K, KpnI, R, EcoRI, Nc, NcoI, No, NotI, S, SfuI. Restriction sites in parentheses were destroyed after the targeting event.

G6Pase gene with a neomycin cassette (Lei et al., 1996) (Fig. 4.1). Exon 3 deletion is predicted to yield a 115-residue truncated G6Pase protein that is enzymatically inactive. Biochemical and molecular analyses demonstrated that, although mature G6Pase mRNA and protein are present in the livers and kidneys of G6Pase$^{+/+}$ and G6Pase$^{+/-}$ mice, they are absent from those of homozygous G6Pase$^{-/-}$ mice (Lei et al., 1996). Therefore, the targeted mutation in exon 3 had resulted in a null *G6Pase* locus in vivo.

5. PHENOTYPE OF THE G6PASE$^{-/-}$ MICE

The G6Pase$^{-/-}$ mice manifest essentially the same phenotype as the human GSD-1a patients, including hypoglycemia, growth retardation, hepatomegaly, kidney enlargement, hyperlipidemia, and hyperuricemia (Lei et al., 1996). In human GSD-1a patients, hepatomegaly is caused primarily by glycogen and lipid storage, and kidney enlargement is due to glycogen storage. Hematoxylin and eosin staining and electron microscopic analyses revealed that the G6Pase$^{-/-}$ mice also have marked glycogen and lipid storage in hepatocytes (Fig. 4.2). These deposits create a uniform mosaic architecture with compression of the liver sinusoids, similar to that found in human GSD-1a patients. There is also marked glycogen accumulation in the tubular epithelial cells of the kidneys, which results in enlargement and compression of the glomeruli (Fig. 4.2). If an effective dietary therapy is not implemented, GSD-1a patients expire early due to severe hypoglycemia and metabolic acidosis secondary to

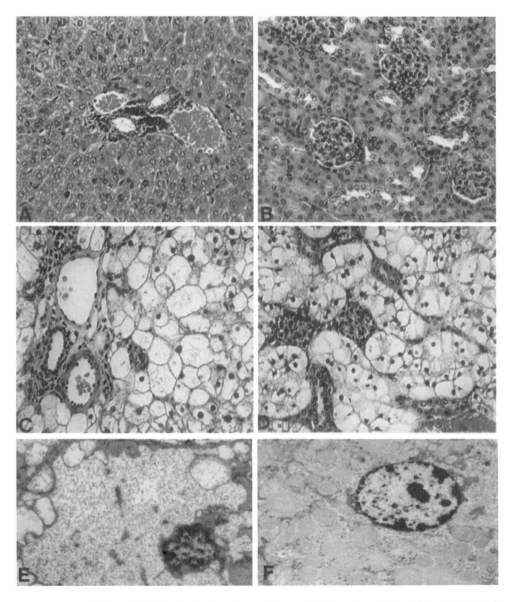

Figure 4.2. Histological analyses of liver and kidney from G6Pase$^{+/+}$ and G6Pase$^{-/-}$ mice. (A–D) Hematoxylin-eosin stained liver and kidney from G6Pase$^{+/+}$ (A, B) and G6Pase$^{-/-}$ (C, D) littermates at 21 days of age. (E, F) Electron micrographs of liver from a G6Pase$^{-/-}$ mouse at 26 days of age. Note the abundance of glycogen particles (E) and lipid droplets (F) in the cytoplasm.

liver G6Pase deficiency and/or end-stage renal disease. Similarly, G6Pase$^{-/-}$ mice die at four to five weeks of age due to severe hepatic and renal lesions. The G6Pase$^{-/-}$ mouse is, therefore, a valid model for studying the pathophysiology of GSD-1a at the cellular and molecular levels and for developing novel therapies for this disorder, including gene therapy.

6. ESTABLISHMENT OF CONDITIONALLY TRANSFORMED HEPATOCYTE LINES

To evaluate methodologies for *G6Pase* gene transfer in vitro as a prelude to gene therapy with the animal model, we immortalized hepatocyte lines from G6Pase$^{-/-}$ mice. These hepatocyte lines also provide in vitro model systems to study the relationship between G6Pase deficiency and biochemical abnormalities observed in GSD-1a and to characterize the G6Pase system.

A frequent problem in establishing differentiated mammalian cell cultures is the overgrowth of these cells by fibroblasts. Therefore, in addition to using a culture condition that supports differentiation of the host cell, primary cells should be infected with a SV40 *tsA* mutant immediately after attachment, preferably within 24 h. For immortalizing hepatocytes, we used culture medium supplemented with 0.1-μm dexamethasone (DEX). Studies have shown that the glucocorticoid hormone is required for liver differentiation in vitro (Chou et al., 1988).

Primary mouse hepatocytes from G6Pase$^{-/-}$, G6Pase$^{+/-}$, or G6Pase$^{+/+}$ littermates are obtained by collagenase digestion of newborn mouse livers from the respective strains and are grown in αMEM supplemented with 10% FBS and 0.1-μm DEX. One day after plating, the cells are washed with medium to remove unattached cells and blood. Then the media are aspirated and cultures are infected at 34°C with a SV40 *tsA* mutant virus (0.5 ml of the virus stock per 25-cm^2 flask) at a MOI of 5–10/cell. After absorption at 34°C for 2–3 h with gentle mixing every 15 to 20 min, the viral fluid is aspirated and the cells are fed with α-MEM containing 4% FBS and 0.1-μm DEX and are incubated at 34°C. The medium (αMEM containing 4% FBS and 0.1-μm DEX) is replaced twice weekly. Clones of transformed cells are identifiable after 4–5 weeks and are isolated and expanded. Transformed hepatocytes are grown and maintained in αMEM containing 4% FBS and 0.1-μm DEX.

Transformed hepatocyte lines are screened by radioimmunoassay for the production of a liver-specific gene product, albumin, because the level of albumin production is often used as an indicator of hepatocyte differentiation. Hepatocyte lines that produced high levels of albumin are further characterized.

The efficiency of transformation by SV40 *tsA* mutants is dependent upon the genetics of the host cell and the age of the animal from which the tissue is obtained. In general, rodent cells are more easily transformed by SV40 *tsA* mutants than the human cells. In addition, the efficiency of transformation of fetal cells by SV40 *tsA* mutants is 10- to 100-fold higher than that of adult cells. To obtain independent clones of transformed cells, it is necessary to serially dilute the infected cultures 16 to 24 h after SV40 transformation. This requirement is especially true for cells of fetal origin.

It is important to recognize that the successful establishment of immortalized cell lines depends upon three important factors: the homogeneity and differentiated state of the starting cells, the ability of the culture condition to sustain host-cell differentiation, and a simple and reliable method to screen cell lines that express tissue-specific products. One should keep in mind that differentiated cell lines, immortalized with a *tsA* mutant of SV40, can de-differentiate following prolonged culture in vitro (Chou, 1989). Thus, cultures should be re-initiated every three to four months to ensure experiments with reproducible results.

7. CHARACTERIZATION OF CONDITIONALLY TRANSFORMED HEPATOCYTE LINES

The SV40 T-antigen is required for maintenance of the transformed state of the cells (Martin and Chou, 1975; Martin, 1981), and cell lines immortalized by a SV40 *tsA* mutant are temperature-sensitive for transformation. However, cell lines with multiple copies of the SV40 genome inserted into the host genome tend to be temperature-resistant (Reviewed in Martin, 1981). To determine if cell lines immortalized with a SV40 *tsA* mutant virus are temperature-sensitive for transformation, we examined the abilities of these cells to overgrow non-transformed cell layers at a permissive and a nonpermissive temperature. We also examined whether these cells express genes that are tissue-specific, re-flecting the differentiated state of the immortalized cells.

8. OVERGROWTH OF NONTRANSFORMED FIBROBLASTS

Human fibroblasts are trypsinized and suspended in αMEM containing 4% FBS at a density of 2×10^5 cells per ml. Each well of a 24-well plate (2 cm^2/well) receives 1.5 ml of the suspension, which produces a confluent mono-layer after attachment. Each well also receives transformed hepatocytes at var-ious concentrations. One plate is incubated at 34°C (permissive temperature), and a duplicate plate is incubated at 39.5°C (nonpermissive temperature). The medium is replaced twice weekly. After two weeks of incubation at either tem-perature, the cells are fixed in absolute methanol and stained with 0.1% Evans blue in phosphate-buffered saline (Martin and Chou, 1975).

All three hepatocyte lines, which are derived from G6Pase$^{-/-}$ (HO-15), G6Pase$^{+/-}$ (HT-5), and G6Pase$^{+/+}$ (W-12) mice, respectively, are reversible for transformation. The results of the overgrowth assay for the G6Pase$^{-/-}$ hepato-cyte line, HO-15, are shown in Fig. 4.3. At 34°C, these cells behaved as trans-formed cells and overgrew the nontransformed cell layers. At 39.5°C, the ability of these cells to overgrow nontransformed cell layers was greatly re-duced, demonstrating that HO-15 cells are temperature-sensitive for mainte-nance of the transformed phenotype.

9. EXPRESSION OF GENES ENCODING G6PASE AND G6PT

Northern-blot hybridization studies were used to examine the ability of the trans-formed hepatocyte lines to express G6Pase and G6PT, two liver-specific genes that are deficient in GSD-1a and GSD-1b, respectively. In the presence of DEX, the ma-ture G6Pase transcript of 2.2 kb was detected in the G6Pase$^{+/+}$ (W-12) and G6Pase$^{+/-}$ (HT-5) hepatocyte lines (Fig. 4.4A). HT-5 cells also expressed low lev-els of an incompletely spliced G6Pase transcript. Studies have shown that G6Pase mRNA expression is enhanced by glucocorticoids and cAMP (Chou et al., 1991). As predicted, G6Pase mRNA expression in W-12 and HT-5 cells was stimulated by the simultaneous addition of DEX and dibutyryl cAMP (Bt$_2$cAMP) (Fig. 4.4A).

The G6Pase$^{-/-}$ hepatocyte line, HO-15, primarily expressed incompletely spliced G6Pase transcripts that had slower electrophoretic mobilities than that

HO-15 Hepatocytes

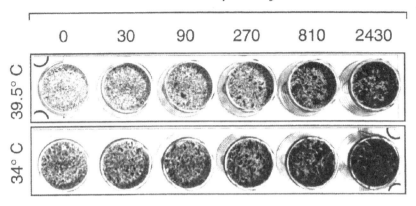

Figure 4.3. Ability of HO-15 G6Pase$^{-/-}$ hepatocytes to overgrow a monolayer of non-transformed human fibroblasts at 34°C and 39.5°C. Each well contains 2×10^5 human fibroblasts and 0, 30, 90, 270, 810, or 2430 of HO-15 cells. Duplicate plates were prepared and incubated either at 34°C or 39.5°C. After two weeks of incubation, cells were fixed and stained.

of the mature transcript (Fig. 4.4A). No mature G6Pase mRNA was expressed by HO-15 hepatocytes in the presence of either DEX or DEX/Bt$_2$cAMP (Fig. 4.4A). The lack of mature G6Pase mRNA expression by HO-15 cells was confirmed by using an exon 3-specific G6Pase probe, which detected a mature G6Pase mRNA of 2.2 kb in W-12 and HT-5 cells but not in HO-15 cells (Fig. 4.4B). It is interesting to note that expression of the immature G6Pase transcripts in HO-15 cells was not stimulated by the simultaneous addition of DEX and Bt$_2$cAMP (Fig. 4.4A).

W-12, HT-5, and HO-15 hepatocyte lines expressed high levels of the G6PT mRNA in the presence of DEX (Fig. 4.5A). Unlike G6Pase, the levels of G6PT mRNA were not augmented by the addition of DEX and Bt$_2$cAMP. At present, little is known about the hormones or factors that regulate G6PT gene expression. These differentiated hepatocyte lines derived from G6Pase$^{-/-}$, G6Pase$^{-/-}$, and G6Pase$^{+/-}$ mice provide an invaluable resource to study the regulation of G6PT gene expression and the interrelationship between G6Pase and G6PT.

10. EXPRESSION OF OTHER LIVER GENES

Tyrosine aminotransferase (TAT) is a liver-specific gene that is normally expressed after birth (Greengard, 1970). The hormonal control of TAT has been one of the most intensively studied subjects in mammalian cell regulation, and TAT gene expression is stimulated by glucocorticoids as well as by cAMP (Ruiz-Bravo and Ernest, 1982). In the presence of DEX, all three hepatocyte lines expressed the TAT mRNA (Fig. 4.5B), albeit at different levels. TAT expression in these mouse hepatocyte lines was further stimulated by the simultaneous addition of DEX and Bt$_2$cAMP, as well as by shifting these cells from a permissive to a nonpermissive temperature (data not shown) as was demon-

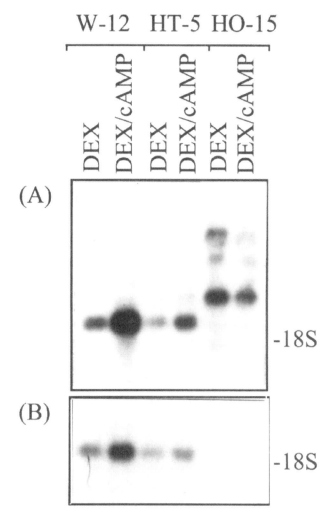

Figure 4.4. Expression of G6Pase mRNA in transformed hepatocyte lines. Poly(A)$^+$ RNAs (2 μg/lane) isolated from W-12 (G6Pase$^{+/+}$), HT-5 (G6Pase$^{+/-}$), and HO-15 (G6Pase$^{-/-}$) cells were separated on formaldehyde-agarose gels and hybridized with a uniformly labeled antisense probe of mouse G6Pase cDNA (A) or exon 3 of the mouse *G6Pase* gene (B).

strated previously in SV40 *tsA* mutant-immortalized hepatocyte lines from wild-type and albino mutant mice (Chou et al., 1991).

Although all three hepatocyte lines were temperature-sensitive for transformation and TAT mRNA expression in these cells was stimulated in cells grown at 39.5°C (differentiated phenotype), we observed little change in G6Pase expression in cells grown at 34°C and 39.5°C (data not shown). It has been reported that the phenotypes of SV40 *tsA* mutant-transformed cells are influenced by several factors (reviewed in Martin, 1981), which include: (1) the multiplicity of infection; (2) the route of infection (virion versus DNA); (3) the genetics of the recipient host cell; and (4) the growth state of the cells immediately after infection.

11. AN IN VITRO MODEL FOR THE EVALUATION OF GENE THERAPY FOR GSD-1A

One of the goals of generating a G6Pase$^{-/-}$ mouse is to develop gene therapies for GSD-1a. The first step in developing a viral gene therapy strategy is to test the ability of the recombinant virus to deliver the transgene in vitro. The HO-15 G6Pase$^{-/-}$ hepatocyte line has been shown to support a *G6Pase* transgene expression directed by a vector based on herpes simplex and adeno-associated viruses (Fraefel et al., 1997). These hepatocytes also supported reporter gene expression directed by liver-specific promoters (Fraefel et al., 1997). Currently, we are developing a recombinant adenovirus-mediated somatic liver gene therapy for GSD-1a and have generated a mouse *G6Pase*-bearing recombinant adenovirus carrying the viral *dlE1/tsE2a* mutations. We have used the HO-15 hepatocytes to test the ability of the recombinant virus to deliver the *G6Pase* transgene. Our results showed that after 24 h of infection with the recombinant mouse G6Pase virus at a MOI of 300 PFU/cell, G6Pase activity in the infected cultures was sevenfold higher than that observed in the mouse liver. Therefore, the G6Pase$^{-/-}$ hepatocytes are a valid in vitro model for developing gene therapy for GSD-1a.

12. CONCLUDING REMARKS

After two decades, the SV40 *tsA* mutant viruses remain a promising agent to immortalize differentiated mammalian cells. These viruses are among the most efficient to immortalize murine cells because mouse is a nonpermissive host for the SV40 virus. Immortalized cell lines retaining tissue-specific functions

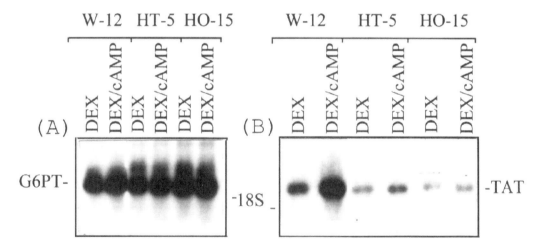

Figure 4.5. Expression of G6PT and TAT mRNA in transformed hepatocyte lines. Poly(A)$^+$ RNAs (2 μg/lane) isolated from W-12 (G6Pase$^{+/+}$), HT-5 (G6Pase$^{+/-}$), and HO-15. (G6Pase$^{-/-}$) cells were separated on formaldehyde-agarose gels and hybridized with a uniformly labeled antisense probe of mouse G6PT (A) or TAT (B).

can be successfully established as long as hormones or other factors that support host cell differentiation are known and a homogeneous population of host cells can be obtained. For a cell type that is difficult to culture in sufficient quantity for transformation, immortalization can still be achieved by introducing, to the transgenic or knockout mice, a SV40 temperature-sensitive T-antigen transgene directed by a tissue-specific promoter.

REFERENCES

Aden DP, Fogel A, Damjanov I, Knowles BB (1979): Controlled synthesis of HBsAg in a differentiated human liver carcinoma-derived cell line. Nature 282:615–616.

Amsterdam A, Zauberman A, Meir G, Pinhasi-Kimhi O, Suh BS, Oren M (1988): Cotransformation of granulosa cells with simian virus 40 and Ha-RAS oncogene generates stable lines capable of induced steroidogenesis. P NAS US 85:7582–7586.

Arion WJ, Lange AJ, Walls HE, Ballas LM (1980): Evidence of the participation of independent translocases for phosphate and glucose-6-phosphate in the microsomal glucose-6-phosphatase system. J Biol Chem 255:10396–10406.

Banks-Schlegel SP, Howley PM (1983): Differentiation of human epidermal cells transformed by SV40. J Cell Biol 96:330–337.

Cairns LA, Crotta S, Minuzzo M, Moroni E, Granucci F, Nicolis S, et al. (1994): Immortalization of multipotent growth-factor dependent hemopoietic progenitors from mice transgenic for GATA-1 driven SV40 tsA58 gene. EMBO J 13:4577–4586.

Chen GL, Halligan LN, Lue NF, Chen WW (1987): Biosynthesis of myelin-associated proteins in simian virus 40 (SV40)-transformed rat schwann cell lines. Brain Res 414:35–48.

Chen Y-T, Burchell A. Glycogen storage diseases. In The Metabolic and Molecular Bases of Inherited Disease, Seventh Edition; Scriver CR, Beaudet AL, Sly WS, Valle D, Eds.; McGrawHill: New York, 1995; pp 935–965.

Chou JY (1978a): Establishment of clonal human placental cells synthesizing human choriogonadotropin. P NAS US 75:1854–1858.

Chou JY (1978b): Human placental cells transformed by tsA mutants of simian virus 40:a model system for the study of placental functions. P NAS US 75:1409–1413.

Chou JY (1983): Temperature-sensitive adult liver cell line dependent on glucocorticoid for differentiation. Mol Cell B 3:1013–1020.

Chou JY (1985): Establishment of rat fetal liver lines and characterization of their metabolic and hormonal properties: use of temperature-sensitive SV40 virus. Meth Enzym 109:385–396.

Chou JY (1989): Differentiated mammalian cell lines immortalized by temperature-sensitive tumor viruses. Mol Endocr 10:1511–1514.

Chou JY, Martin RG (1974): Complementation analysis of simian virus 40 mutants. J Virology 13:1101–1109.

Chou JY, Ruppert S, Shelly LL, Pan C-J (1991): Isolation and characterization of mouse hepatocyte lines carrying a lethal albino deletion. J Biol Chem 266:5716–5722.

Chou JY, Schlegel-Haueter SE (1981): Study of liver differentiation in vitro. J Cell Biol 89:216–222.

Chou JY, Wan Y-JY, Sakiyama T (1988): Regulation of rat liver maturation in vitro by glucocorticoids. Mol Cell B 8:203–209.

Cluzeaud F, Bens M, Wu MS, Li Z, Vicart P, Paulin D, et al. (1996): Relationships between intermediate filaments and cell-specific functions in renal cell lines derived from transgenic mice harboring the temperature-sensitive T antigen. J Cell Phys 167:22–25.

Fong HKW, Chick WL, Sato GH (1981): Hormones and factors that stimulate growth of a rat islet tumor cell line in serum-free medium. Diabetes 30:1022–1028.

Fraefel C, Jacoby DR, Lage C, Hilderbrand H, Chou JY, Alt FW, et al. (1997): Gene transfer into hepatocytes mediated by helper virus-free HSV/AAV hybrid vector. Mol Med 3:813–825.

Gao Q, Hauser SH, Liu XL, Wazer DE, Madoc-Jones H, Brand V (1996): Mutant p53-induced immortalization of primary human mammary epithelial cells. Canc Res 56:3129–3133.

Georgoff I, Scott T, Isom HC (1984): Effect of simian virus 40 on albumin production by hepatocytes cultures in chemically defined medium and plated on collagen and non-collagen attachment surfaces. J Biol Chem 259:9595–9602.

Gey GO, Coffman WD, Kubicek MT (1952): Tissue culture studies of the proliferative capacity of cervical carcinoma and normal epithelium. Canc Res 12:264–265.

Greengard O. The developmental formation of enzyme in rat liver. In Mechanisms of Hormone Action; Litwack IG, Ed.; Academic Press: New York, 1970; pp 53–85. He X, Frank DP, Tabak LA (1990): Establishment and characterization of 12S adenoviral E1A immortalized rat submandibular gland epithelial cells. Bioc Biop R 179:336–342.

Hiraiwa H, Pan C-J, Lin B, Moses SW, Chou JY (1999): Inactivation of the glucose-6-phosphate transporter causes glycogen storage disease type 1b. J Biol Chem 274:5532–5536.

Isom HC, Tevethia J, Taylor JM (1980): Transformation of isolated rat hepatocytes with simian virus 40. J Cell Biol 85:651–659.

Kiyono T, Foster SA, Koop JI, McDougall JK, Galloway DA, Klingelhutz AJ (1998): Both Rb/p16INK4a inactivation and telomerase activity are required to immortalize human epithelial cells. Nature 396:84–88.

Lei K-J, Chen Y-T, Pan C-J, Ward JM, Mosinger B, Lee EJ, et al. (1996): Glucose-6-phosphatase dependent substrate transport in the glycogen storage disease type 1a mouse. Nat Genet 13:203–209.

Lei K-J, Shelly LL, Pan C-J, Sidbury JB, Chou JY (1993): Mutations in the glucose-6-phosphatase gene that cause glycogen storage disease type 1a. Science 262:580–583.

Li WI, Chen CL, Chou JY (1989): Characterization of a temperature-sensitive β-endorphin secreting transformed endometrial cell line. Endocrinology 125:2862–2867.

Martin RG (1981): The transformation of cell growth and transmogrification of DNA synthesis by simian virus 40. Adv Canc R 34:1–68.

Martin RG, Chou JY (1975): Simian virus 40 functions required for the establishment and maintenance of malignant transformation. J Virology 15:599–612.

Nordlie RC, Sukalski KA. Multifunctional glucose-6-phosphatase: a critical review. In The Enzymes of Biological Membranes, Second Edition; Martonosi AN, Ed.; Plenum Publishing: New York, 1985; pp 349–398.

Pan C-J, Lei K-J, Annabi B, Hemrika W, Chou JY (1998): Transmembrane topology of glucose-6-phosphatase. J Biol Chem 273:6144–6148.

Pattillo RA, Gey GO (1968): The establishment of a cell line of human hormone-synthesizing trophoblastic cells in vitro. Canc R 28:1231–1236.

Reid LM, Jefferson DM. Cell culture studies using extracts of extracellular matrix to study growth and differentiation in mammalian cells. In Mammalian Cell Culture; Mather JP, Ed.; Plenum Publishing: New York, 1984; pp 239–280.

Rhim JS, Jay G, Arnstein P, Price FM, Sanford KK, Aaronson SA (1985): Neoplastic transformation of human epidermal keratinocytes by AD12-SV40 and Kirsten sarcoma viruses. Science 227:1250–1252.

Ruiz-Bravo N, Ernest MJ (1982): Induction of tyrosine aminotransferase mRNA by glucocorticoids and cAMP in fetal rat liver. P NAS US 79:365–368.

Sirvastava RK, Zilberstein M, Ou JS, Mayo KE, Chou JY, Gibori G (1995): Development and characterization of a SV-40 transformed temperature sensitive rat antimesometrial decidual cell line. Endocrinology 136:1913–1919.

Stampfer MR, Bartley JC (1985): Induction of transformation and continuous cell lines from normal mammary epithelial cells after exposure to benzo[α]pyrene. P NAS US 82:2394–2398.

Takayama H, Tanigawa T, Tanaka Y, Kimura G (1986): Induction of fibronectin expression, actin cable formation, and entry into S phase following reexpression of T- antigen in mouse macrophages transformed by the tsA640 mutant of SV40. J Cell Phys 128:271–278.

Tooze J. DNA Tumor Viruses; Cold Spring Harbor Laboratory: New York, 1981.

Vitry F, Camier M, Czernichow P, Benda P, Cohen P, Tixier-Vidal A (1974): Establishment of a clone of mouse hypothalamic neurosecretory cells synthesizing neurophysin and vasopressin. P NAS US 71:3575–3579.

Whitehead RH, VanEeden PE, Noble MD, Ataliotis P, Jat PS (1993): Establishment of conditionally immortalized epithelial cell lines from both colon and small intestine of adult H-2K^b-tsA58 transgenic mice. P NAS US 90:587–591.

Yankaskas JR, Haizlip JE, Conrad M, Koval D, Lazarowski E, Paradiso AM, et al. (1993): Papilloma virus immortalized tracheal epithelial cells retain a well-differentiated phenotype. Am J Physiol 264:1219–1230.

Yoakum GH, Lechner JF, Gabrielson EW, Korba BE, Malan-Shibley L, Willwy JC, et al. (1985): Transformation of human bronchial epithelial cells transfected by Harvey ras oncogene. Science 227:1174–1179.

Zaret KS, DiPersio CM, Jackson DA, Montigny WJ, Weinstat DL (1988): Conditional enhancement of liver-specific gene transcription. P NAS US 85:9076–9080.

Zilberstein M, Chou JY, Lowe Jr. WL, Shen-Orr Z, Roberts CT, Leroith D, et al. (1989): Expression of insulin growth factor-I and its receptor by SV40-transformed rat granulosa cells. Mol Endocr 3:1488–1497.

CHAPTER 5

THE VP16-DEPENDENT BINARY SYSTEM FOR INDUCIBLE GENE EXPRESSION IN TRANSGENIC MICE

CLAUDIA KAPPEN

1. INTRODUCTION

Transgenic mice are powerful tools for the study of gene function and provide valuable disease models. However, when the gene of interest interferes with the normal developmental process, the usefulness of conventional transgenic approaches is limited (for reviews, see Hanahan, 1989; Westphal and Gruss, 1989; Byrne et al., 1991). Frequently in such cases, ectopic or high-level expression of the transgene causes embryonic, peri- or post-natal death. As a result, establishing permanent transgenic mouse lines expressing these types of genes has proven difficult. Binary transgenic mouse systems were designed to circumvent these difficulties. This article will review the general concept and engineering principles of the VP16-dependent binary system and its applications to the study of the function of homeobox genes in embryonic development and cell differentiation.

Hox genes encode transcription factors that are expressed in specific spatial and temporal patterns in the developing embryo. It has been amply demonstrated that the loss-of-function of Hox genes results in malformations and patterning defects, particularly in the axial and appendicular skeleton (McGinnis and Krumlauf, 1992; Horan et al., 1995; Capecchi, 1996; Chen and Capecchi, 1997; Rijli and Chambon, 1997). However, it is not well understood how Hox transcription factors act at the cellular level, not least because specific target genes remain to be identified. A particular problem in investigating this aspect is presented by the fact that in any given region or cell in the developing skeleton, the expression patterns of several Hox genes overlap. Thus, the specific complement as well as the overall level of Hox gene expression is likely to

Genetic Manipulation of Receptor Expression and Function,
Edited by Domenico Accili.
ISBN 0-471-35057-5   Copyright © 2000 Wiley-Liss, Inc.

influence cell differentiation and/or proliferation in a given region of the embryo (Duboule, 1995; Capecchi, 1996). To find out how Hox transcription factors control these cellular processes, one would have to eliminate all Hox genes specifically from one region or tissue, a task that has been and continues to be technically very difficult. In light of this condition, we chose a transgenic approach in which the expression of only one of the genes would be manipulated. Our reasoning was that the overexpression of a Hox gene in those cells where it is normally expressed might amplify its normal function in these cells and thus allow us to assess the cellular consequences of Hox gene regulation. However, conventional Hox-transgenes are typically detrimental to development, resulting in pre-, peri- and post-natal death (Balling et al., 1989; Jegalian and DeRobertis, 1992; Kaur et al., 1992; McLain et al., 1992; Pollock et al., 1992; Charite et al., 1994; Zhang et al., 1994). Thus, it was necessary to create mice in which the expression of Hox transgenes is inducible in a conditional fashion by transcriptional transactivation.

2. THE VP16-DEPENDENT BINARY TRANSGENIC MOUSE SYSTEM

The VP16-dependent binary system is a two-tiered gene-activation system (Byrne and Ruddle, 1989) in which expression of the transgene of interest is induced by the potent viral transactivator VP16. VP16 is an integral coat protein of Herpes Simplex Virus (HSV) and is used by the virus for immediate early gene transcription upon cell infection. Genes transactivated by VP16 (immediate early genes, or IE genes) harbor recognition sequences for VP16 transactivation; the common motif being TAATGARAT (McKnight et al., 1987; Preston et al., 1988). Thus, any gene that is artificially linked to an IE-promoter is expected to be transactivated in the presence of VP16. In this way, transgenes that are otherwise detrimental to development can be inherited in silent form, and their expression occurs only in offspring that carry transgenes encoding both components of the binary system: the transactivator transgene encoding VP16 (TA) and the transresponder transgene (TR), which encodes the gene of interest (Fig. 5.1). The outcome of transresponder gene expression may still be detrimental to the survival of an individual that inherits both transgenes. However, the existence of parental transgenic mouse strains with genetically defined stable transgene loci allows the reproducible creation—simply by breeding—of an unlimited supply of double transgenic progeny.

There are a number of criteria that such a binary system needs to satisfy to be useful (Byrne et al., 1991) for investigating gene function in embryonic development: (1) VP16-mediated transactivation has to work in the embryo in multiple tissues and throughout development. (2) Expression of the TA transgene needs to be specific and at satisfactory level. (3) The TR transgene needs to be silent; "leaky" or basal expression of the TR transgene is undesirable. (4) The transactivator itself should not interfere with development. (5) The system should achieve or exceed the efficiency of conventional transgenic technology. Inherent in any binary transgenic or recombination-based system are additional features not typically afforded by conventional technology (Rossant and

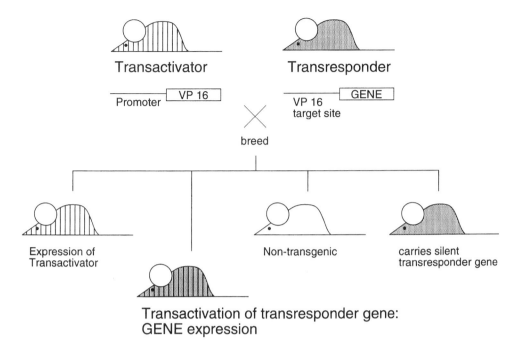

Figure 5.1. VP16-dependent transactivation in transgenic mice. The VP16-dependent binary system for inducible gene expression in transgenic mice consists basically of two components, the transactivator and the transresponder. One transgenic line (transactivator, TA, left), contains the VP16 gene of HSV-1 under the control of regulatory elements of choice. The other transgenic line (transresponder, TR, right) harbors the transgene to be expressed linked to the IE promoter of HSV-1. When parental mice heterozygous for each of the transgene insertions are crossed, four genotypes are possible in the progeny. Offspring will inherit only the TA transgene (far left), only the TR transgene (far right), neither (center right) or both (center left) of the transgenes. In those cells expressing VP16 protein, transactivation of the IE-promoter occurs, resulting in expression of the TR gene.

McMahon, 1999), such as (6) the possibility to control the level of transactivation and effectively create an allelic series and (7) the versatility of comparative and combinatorial transgenesis. In the following, I will discuss how these engineering aspects have been addressed with the VP16 system, as a paradigm of binary systems.

VP16-Mediated Transactivation in the Developing Embryo: Temporal Regulation and Tissue-Specificity

Byrne and Ruddle (1989) originally demonstrated that an IE-CAT TR transgene could be activated by VP16. VP16 was either supplied by infection with HSV or by a transgene in which the VP16 gene was linked to the promoter from the murine neurofilament light chain gene. In both cases, transactivation could be detected in HSV-infected animals or in progeny transgenic for both the TA and TR transgenes. The results showed a strict dependence of CAT activity on the presence of VP16 in mice at 6 to 8 weeks of age. These experiments provided proof-of-concept evidence that VP16-mediated trans-activation can be used to induce transgene expression. However, to make the system applicable to the purpose of studying Hox gene function in development, it was necessary to establish whether VP16 could also be used to trans-activate transgenes in developing embryos and in which tissues the system would be functional.

We chose a promoter/enhancer that was known to be developmentally regulated during embryogenesis, the 5' upstream fragment from the murine Hoxc-8 gene. This promoter is specifically active in the posterior region of the developing embryo (Bieberich et al., 1990; Gardner et al., 1996) so that anterior tissues would serve as negative control within the same embryo. We generated TA mice in which the expression of VP16 was controlled by the Hoxc-8 promoter. To assay for VP16 transactivation activity at the cellular level, we created TR mice that harbored an IE-LacZ reporter transgene. This approach allowed us to visually detect the expression of the TR transgene product by virtue of β-galactosidase staining in embryonic tissues (Gardner et al., 1996). Prior experiments had shown that a conventional Hoxc-8-LacZ transgene was expressed in the posterior region of the developing embryo, in the spinal cord, somites, and lateral mesoderm (Bieberich et al., 1990). We found essentially the same patterns of LacZ expression in embryos derived through the VP16 binary system, in which they inherited both the Hoxc-8-VP16 TA and the IE-LacZ TR transgenes. There was no staining in anterior regions of the embryos at 9.5 days and little β-galactosidase activity at 10.5 days of development (Gardner et al., 1996). Furthermore, the spatial boundaries and tissue distribution of staining reproduced the pattern of activity expected from the Hoxc-8 promoter. Thus, VP16-mediated TR gene activation faithfully reflected the specificity of the promoter controlling VP16 expression.

By using our IE-LacZ TR mice, we could specifically investigate which tissues were permissive for VP16 transactivation during embryogenesis. As described above, the Hoxc-8-VP16 transactivator reproduced the activity of the endogeneous Hoxc-8 promoter in neuroectodermally and mesodermally derived tissues, including skeletal elements or, for example, the skin. At later stages of development, we have found the Hoxc-8 promoter to be active in chondrocytes (Yueh et al., 1998; Cormier-Regard and Kappen, unpublished observations). The permissiveness of neural tissues for VP16 transactivation was demonstrated by Byrne and Ruddle's original observations for adult brain (1989) and by our own studies using the mouse neurofilament light chain gene promoter to drive VP16 expression (Yaworsky et al., 1997). In these embryos, transient activity of the NF-promoter was also found in embryonic skeletal muscle (Yaworsky et al., 1997), indicating that muscle tissues can support VP16-dependent transactivation. We further assayed β-galactosidase activity from the IE-LacZ TR transgene by using various TA mouse strains and VP16 transgene insertion loci (unpublished observations). Figure 5.2 shows representative examples of embryonic tissues in which staining was apparent. In addition to the developing skeleton (Yueh et al., 1998), muscle, and nervous system (Yaworsky et al., 1997), VP16-mediated transactivation was detected in salivary gland, mesenchyme around the eye, in the lens and the retina

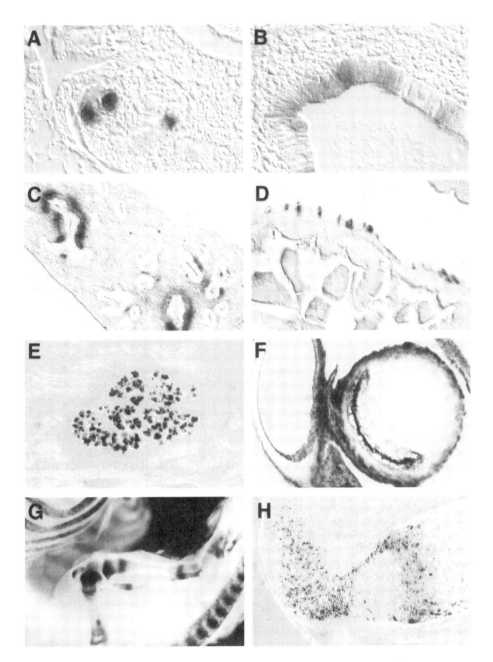

Figure 5.2. VP16-mediated transactivation of an IE-LacZ TR gene in various tissues during development.Embryos were isolated from timed matings of transactivators to IE-LacZ TR mice at 13.5 (A, B), 16.5 (C) and 17.5 (D–H) days post-coitum and processed for β-galatosidase staining. Specimens were photographed as whole mounts, and Paraffin-embedded specimens were sectioned at 10 μm thickness. Dorsal is up and anterior is left for A-F and H. Expression of β-galatosidase activity is detectable in A) mesonephric tubules of the developing ovary; B) mucosal lining of the stomach; C) bronchioles of the lung; D) mucosal lining and sympathetic innervation of the small intestine; E) salivary gland; F) outer neuronal layer of eye and subjacent sclera; G) cartilage of costals, long bones, and vertebra; and H) growth zones and cartilage of the knee joint. (See color plates.)

(Yaworsky et al., 1997), skin, lung, stomach, intestine, kidney, urogenital tract, and notochord. Faint staining was also apparent in a few cells in the liver. Taken together, these results indicate that most tissues are permissive for VP16-dependent transactivation. The efficacy of VP16 transactivation conceivably is not only influenced by the level of VP16 expression (Rundle et al., 1998) but also by the stability of the protein and availability of its interacting partners. There have not been any studies that would shed light on the stability or turnover of VP16 in vivo or in infected cells. However, we have observed that tissues that transiently expressed VP16 (as measured by transactivation of the IE-LacZ TR gene and X-gal staining) would not show evidence of transactivation one day later in development (Yaworsky et al., 1997; Cormier-Regard and Kappen, unpublished observations). We thus conclude that VP16 is only moderately stable, and that it can be eliminated from cells that no longer transcribe the TA transgene within one day.

The existence of mouse lines with genetically defined stable transgene loci for the Hoxc-8-VP16 transactivator also enabled us to unequivocally determine the temporal onset of transcriptional activity from the Hoxc-8 promoter. VP16 activity, as measured by X-gal staining, was detectable at day 8 of development, indicating that the promoter becomes active shortly before this time (Gardner et al., 1996). In an analogous approach, we determined the temporal onset of activity of murine neurofilament promoter by day 8.5 post-coitum (Yaworsky et al., 1997). The collective data show that VP16-mediated transactivation can be achieved in transgenic embryos as early as day 8 in post-implantation development. In a separate study (see below), we also showed that introduction of VP16 transgenes could result in transactivation of the IE-LacZ TR gene in pre-implantation embryos from the two-cell to the blastocyst stage (Yueh et al., 2000; Mol. Reprod. Dev., in press). Taken together, our results demonstrate that VP16-mediated transactivation of transgenes is feasible during the earlier as well as later stages of embryogenesis.

Functional Expression of the TA and Effective Silence of the TR Transgene

These two questions are conceptually linked in that they address technical issues that are crucial for the engineering of any binary system, including those based upon repression or recombination. If, for example, the transactivator is only weakly expressed, the TR gene may not be activated at sufficient levels. In addition, ectopic expression of the transactivator is undesirable, as would be uninduced expression of the TR transgene. The potential presence of VP16 in ectopic sites was easily detectable in our system by crossing transactivators to IE-LacZ TR mice. For the Hoxc-8 promoter (9 independent lines tested; Gardner et al., 1996 and unpublished observations) and the neurofilament promoter (6 independent lines tested; Yaworsky et al., 1997), we did not detect major ectopic activities with the exception of one Hoxc-8-VP16 insertion site that resulted in widespread VP16 activity. Thus, as with conventional transgenic technology, the specificity of VP16 expression is largely directed by the respective promoter. However, we did notice that embryos inheriting NF-VP16 TA transgenes required a longer time to reach comparable β-galactosidase staining intensity than animals in which the transactivator was controlled by the enhancer from the rat nestin gene (Yaworsky and Kappen, unpublished observations). This finding seems to indicate that the 'strength' of promoters/enhancers may influence the levels of VP16 expression in TA strains. Because we were not able to detect the transactivator at the protein level (unpublished results), we used β-galactosidase enzyme activity levels to assay for the level of VP16 expression. In this way, we showed that different TA lines reproducibly expressed different levels of VP16 activity, ranging from about 1.5- to 25-fold over background basal levels (Rundle et al., 1998).

An important question has been raised regarding the "leakiness" of the promoter to which the TR transgene is linked. If this transgene would be expressed, it by itself could already induce developmental abnormalities similar to conventional transgenes. Furthermore, "silent" transgenes can act as promoter or enhancer traps, effectively being activated by genomic enhancers in the absence of strong activity of their own. This possibility needs to be ruled out before a given TR strain can be used effectively. Byrne and Ruddle (1989) did not detect CAT activity in the absence of VP16, indicating that the IE-promoter was silent in adult mice. Analogously, we did not observe staining in IE-LacZ transgenic mice in the absence of VP16 (Gardner et al., 1996; Yaworsky et al., 1997; and unpublished results) with the exception of one TR line. In this line (called Z16), the insertion site of the transgene on chromosome 12 renders it transiently expressed in the apical ectodermal ridges of the developing limbs in the absence of VP16 (Gardner and Kappen, 2000; J. Exp. 2001, in press). In contrast to our experience, a report by Mitchell showed β-galactosidase staining in the absence of VP16 in multiple organs of transgenic mice that carried the LacZ gene fused to an IE-promoter (1995). His transgenes included sequences from the mouse protamin gene, which might confer enhancer activity upon an otherwise silent transgene. One additional technical difference to our experiments was a long overnight incubation of the tissues in X-gal (Mitchell, 1995), whereas we routinely terminated our reactions after 2 hr. To resolve the question of "leakiness", we chose three experimental approaches: (1) β-galactosidase staining assay for potential expression of IE-LacZ transgenes; (2) RT-PCR for TR gene expression in independent lines; (3) evaluation of TR transgenic strains for signs of functional TR transgene expression.

To systematically investigate the activity of the IE-promoter during development, we generated and tested 10 lines of IE-LacZ TR mice (Gardner et al., 1996). In none of these did we observe any trace of staining in the embryo; a few lines were also tested at 3, 6, and 9 months of age, at which time neither they nor the Z16 line exhibited any appreciable β-galactosidase activity. Furthermore, there was no X-gal staining in 7 independent transient IE-LacZ transgenic embryos at 10.5 days of development. Taken together, our staining data on IE-LacZ transgenic mice strongly suggest that the IE-promoter is indeed silent. When we used RT-PCR, we did detect evidence for the presence of mRNA produced from the transreponder transgene at high PCR cycle numbers (Rundle et al., 1998). However, the specific bands were generally faint in the absence of VP16, whereas the signal intensity was increased significantly in the presence of VP16 (Rundle et al., 1998). Furthermore, we did not observe any developmental defects in IE-Hoxc-8 or IE-Isl-1 TR lines that were made homozygous for their respective transgene locus, arguing against appreciable "leakiness" of the IE-promoter in these strains. Collectively, our results demonstrate, by three independent criteria, that the IE-promoter is transcriptionally inactive in transgenic mice in the absence of VP16.

Potential Teratogenicity or Toxicity of the Transactivator

The possibility of transactivator toxicity has long been raised as a caveat for any binary system (Byrne and Ruddle, 1989; Rossant and McMahon, 1999). Clearly, if VP16 itself significantly affects embryonic development or interferes with normal cellular functions, its usefulness as a transactivator would be extremely limited. To systematically investigate this possibility, we microinjected a DNA fragment in which the VP16 gene was linked to the human β-actin promoter (Gunning et al., 1987). Out of 130 founder animals screened, only 8 were transgenic (Yueh et al., 2000; Mol. Reprod. Dev., in press). From those, five animals were crossed in an attempt to establish transgenic lines. However, we found much reduced transmission ratios for these transgene loci and only two were transmitted in the second generation. Moreover, none of these transgenes were able to induce IE-LacZ TR gene expression in embryos at 10.5 days of development (Yueh et al., 2000; Mol. Reprod. Dev., in press). These data suggested to us that the few surviving transgenic animals probably carried the transgene in silenced form and that the expression of VP16 from the β-actin promoter likely had detrimental effects on development.

To determine if widespread VP16 expression was toxic to embryos, we injected the β-actin-VP16 transgene directly into IE-LacZ transgenic fertilized eggs. This approach allowed us to assay for embryo survival and VP16-mediated transactivation, simultaneously. Out of 157 transient embryos tested, only 5 were transgenic, but they did not exhibit staining at 10.5 days of development (Yueh et al., 2000; Mol. Reprod. Dev., in press). Thus, any surviving embryos were negative for VP16 expression, suggesting that VP16 induced embryonic lethality before day 10.5. By injecting various VP16 transgene constructs into IE-LacZ transgenic fertilized eggs, we then assayed whether VP16 may interfere with pre-implantation development. This was indeed the case, as VP16 expression from the β-actin-promoter transgene significantly reduced survival of embryos to the blastocyst stage in culture. Any embryos that expressed VP16, as determined by β-galactosidase staining, were retarded in their growth and seemed to deteriorate before ever reaching the mature blastocyst stage. In contrast, the survival rate for embryos injected with a Hoxc-8-VP16 transgene and other control DNA fragments (Yueh et al., 2000; Mol. Reprod. Dev., in press) was the same as for un-manipulated embryos. We concluded from these experiments that VP16 expression at high levels is detrimental to pre-implantation embryo development.

A possible explanation for the lethality at early stages is that VP16, through binding to cellular proteins, effectively diminishes the availability of these factors for essential cellular processes, such as replication, transcription, or proliferation. Binding to VP16 has been demonstrated for replication factor A (RFA; He et al., 1993), a single-stranded DNA-binding protein required for DNA replication, repair, and recombination (reviewed in Wold, 1997), the RNA polymerase II holoenzyme (Hengartner et al., 1995), TFIIA (Kobayashi et al., 1995), TFIIB (Lin and Green, 1991), and TFIID (Stringer et al., 1990; Ingles et al., 1991). Whereas the human POU-domain transcription factor Oct-1 (Gerster and Roeder, 1988; Kristie et al., 1989; Stern et al., 1989) is known to bind VP16, this activity is controversial for mouse Oct-1 (Suzuki et al., 1993). Indeed, it is unlikely that interference with Oct-1 would account for VP16 toxicity, as the disruption of the murine Oct-1 gene itself induces lethality only at or after birth (Ohno et al., 1993; Koyasu et al., 1994). VP16 is imported into the nucleus (Boissiere et al., 1999) and recruited into the transactivation complex by host cell factor (HCF, C1), a family of proteins involved in cell proliferation (Wilson et al., 1993). Detrimental effects of VP16 may thus be mediated directly through HCF or other interacting proteins, such as leucine-zipper transcription factors (Freiman and Herr, 1997; Lu et al., 1998). Whatever the mechanistic basis for VP16 toxicity, our results serve as a caveat for any approach using the VP16 transactivator or portions thereof in very early embryos.

However, our experience with generating TA mice illustrates that toxicity should be less of a concern at later stages of development or in the adult. We were able to establish transgenic lines with the Hoxc-8-VP16, NF-VP16, and nestin-VP16 transgenes (Gardner et al., 1996; Yaworsky et al., 1997; Yueh et al., 1998; Yaworsky et al., in preparation), in which VP16 expression is evident from day 8 on. Thus, it appears that VP16 is well-tolerated after gastrulation. It is still conceivable that specific cell types in later embryogenesis or in the adult mouse may be especially susceptible to detrimental effects of VP16, but in the absence of promoters—enhancers with identical transcriptional activity in different cell types—this possibility is difficult to investigate.

Efficiency of VP16-Mediated Transactivation in Transgenic Mice

To compare the efficiency of the VP16-dependent transactivation system to conventional transgenic approaches, we tested several Hoxc-8-VP16 TA strains in crosses to the same IE-LacZ TR strain. Out of 7 independent TA strains, one harbored a silent integration on the Y-chromosome (Gardner and Kappen, 1998), one displayed ectopic in addition to specific transactivator activity, and the remaining 5 faithfully reflected the activity of the Hoxc-8 promoter. These findings represent a ratio of 71% (5 out of 7) transactivators with appropriate VP16 expression. To assay efficacy on the transresponder side, we determined the number of independent TR strains in which we could achieve transactivation by the same transactivator. Our results with >20 TR lines (Rundle et al., 1998; and unpublished observations) indicate efficient transactivation for at least 70% of the tested TR lines. These results are comparable with the efficiency of expression of conventional transgenes (Hanahan, 1989). The occurrence of phenotypic alterations as a result of transgene activation also appeared to be comparable, as 2 out of 3 IE-Hoxc-8 and 3 out of 4 IE-Isl-1 TR strains produced visible abnormalities (Yueh et al., 1998; Muller et al., manuscript in preparation). Taken together, these results indicate that the VP16-dependent binary system achieves the same efficacy as conventional transgenic technology.

Control of the Level of Transactivation and Allelic Series

The level of VP16 expression is largely controlled through the respective promoter, as well as by transgene copy number and integration sites. Thus, different lines of TA mice can represent an allelic series with different levels of TR gene activation when bred to the same TR strain. Genetic approaches for controlling transactivator activity use breeding of animals to homozygosity for the TA, or the TR locus, or for both as depicted in Figure 5.3. In this way, we could demonstrate that increases in transgene dosage will increase β-galactosidase activity, at least as measured in extracts from 9.5-day p.c. embryos (Rundle et

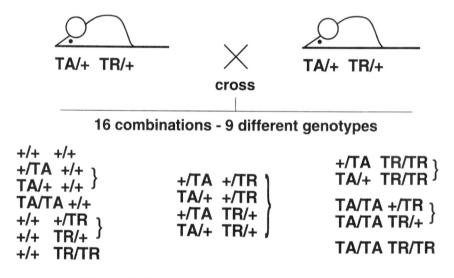

Figure 5.3. Regulation of transgene expression by gene dosage. Crosses of two mice heterozygous for two unlinked markers (TA and TR, presumably intergrated on different chromosomes) formally give rise to 16 different combinations, resulting in 9 different genotypes in inbred mice (16 genotypes in hybrids). Progeny with genotypes listed on the left inherit none or only one of the transgenes in hemizygous or homozygous form. Progeny with genotypes listed in the middle effectively resemble the transgene status of their parents (although the genetic background would be different in hybrids). Progeny with genotypes listed on the right inherit both TA and TR transgenes in excess of hemizygosity. The expected frequency of these genotypes can be increased in crosses of a double hemizygous parent to homozygotes for either TA or TR transgene (TA/+ TR/+ x TA/TA +/+ or TA/+ TR/+ x +/+ TR/TR; see Methods section).

al., 1998). It is interesting to note that homozygosity at the TA locus was more effective in increasing enzyme activity than homozygosity at the TR locus. These results suggest that the levels of TR transgene expression are probably not limited by copy number or availability of the TR gene locus. Analogously, the severity of developmental defects was stronger when the TA rather than the TR locus was present in homozygous form (Yueh et al., 1998; see below). Thus, superimposition of transgene loci provides a defined genetic approach to the control of transgene expression levels.

Combinatorial and Comparative Transgenesis

Another approach for increasing transactivation would be to cross independent TA transgenes into the same mouse and thereby increase gene dosage. Although we do not have direct evidence at the moment, the value of such experiments is dependent on the time and effort it would take to combine and maintain independent transgenes in the same strain. The same condition applies to combinations of TR transgenes. In fact, the simultaneous presence of the IE-LacZ reporter and the TR transgenes enables a direct inquiry into the fate and phenotype of cells in which the TR transgene is activated. This situation can either be accomplished by incorporating a reporter gene as a fusion into the TR locus or by breeding. Such experiments are currently underway in my laboratory.

3. APPLICATION OF THE VP16-DEPENDENT BINARY SYSTEM TO STUDY THE FUNCTION OF HOMEOBOX GENES

Our laboratory developed the VP16-dependent binary system to investigate the functional activities of homeodomain transcription factors in vivo. The long-term goal of these studies is to define the cellular and molecular mechanisms by which transcriptional regulators in the embryo control developmental and cell differentiation pathways. The specific advantage of the binary system is that it enables us to establish stable parental transgenic mouse lines from which we can reproducibly derive genetically defined Hox transgenic mice. Our specific aims are: (1) to characterize the cellular and molecular consequences resulting from ectopic or overexpression of Hox transgenes, including the developmental pathogenesis of phenotypes; (2) to employ rigorous controls in which another transgene is activated in identical fashion; (3) to perform comparative studies on the role of the same Hox gene in different tissues; (4) to isolate cell populations from genetically defined animals; (5) to use the comparison of transgenic with control mice to isolate differentially regulated genes; and (6) to investigate the role of such genes in the manifestation of the developmental defects induced by misexpression of Hox genes. Here I summarize our progress in applying comparative and combinatorial transgenesis in the VP16-dependent binary system to study homeobox gene function.

Transactivation of Hoxc-8: The Severity of Developmental Defects is Dosage-Dependent

Using the Hoxc-8-VP16 TA and IE-Hoxc-8 TR mice, we activated Hoxc-8 expression under control of its own promoter. Our expectation was that this approach would induce abnormalities in patterning of the skeletal system as was observed in conventional Hoxc-8 transgenic mice (Pollock et al., 1992, 1995), where its expression was driven by the heterologous promoter from the Hoxa-4 gene. In addition, it was possible that other tissues could be affected in which the Hoxc-8 promoter was active during development, such as, for example, the spinal cord in the thoracic region and more posterior, the dorsal root ganglia, the skin, or the kidney (Gardner et al., 1996; Yueh et al., 1998; and unpublished observations).

The double transgenic mice generated in crosses of Hoxc-8-VP16 TA to IE-Hoxc-8 TR were all born with open eyes and subsequently developed eye lesions (Yueh et al., 1998). This outcome indicated that the IE-Hoxc-8 TR transgene was being activated. Yet, the animals did not exhibit other gross defects and were apparently healthy and fertile. This condition allowed us to increase TR gene expression by increasing transgene dosage. The general scheme for this approach is shown in Figure 5.3.

When combining transgene loci in excess of hemizygosity (TA/+ TR/TR; TA/TA TR/+; TA/TA TR/TR), we found that no progeny with these genotypes survived (Yueh et al., 1998). Instead, they died shortly during or after birth. Those that were born alive attempted to breathe but ultimately succumbed to pulmonary failure. The animals that were born dead had the double homozygous genotype. These results indicated that increased transgene dosage elevated TR gene expression and the severity of developmental defects. Thus, by superimposition of transgene loci, we could demonstrate that the effects of Hoxc-8 overexpression on phenotype were dosage-dependent.

Consequences of Hoxc-8 Overexpression: Defects in Cartilage Differentiation

Analysis of the axial skeleton in the newborn Hoxc-8 transgenic mice revealed substantial abnormalities in the rib cartilages and morphogenetic defects in vertebrae (Yueh et al., 1998). In fact, the rib cartilage was very soft, suggesting structural defects as the cause of problems in lung inflation. Furthermore, in the double homozygous transgenic skeleton, the whole rib cage was distorted, reflecting the lack of tensile strength and rigidity of thoracic cartilages. There was a significant reduction in alcian blue staining of cartilaginous structures, and this reduction increased with gene dosage. Thus, the structural deficiencies likely were accounted for by a progressive decrease in proteoglycan content of the ribs, for example. The morphogenetic abnormalities in vertebral structures included absence of neural arch fusion and absence of articulation of the neural arches with the central body of vertebrae; in addition, intervertebral cartilages were weak or missing. This phenotype is characterized by insufficient differentiation of cartilage and delayed maturation of skeletal elements (Yueh et al., 1998). In contrast, patterning defects were reported for other Hox gene transgenic mice (Jegalian and DeRobertis, 1992; Pollock et al., 1992; Charite et al., 1994; Pollock et al., 1995), in which transgene expression was driven by promoters with activity in regions ectopic to the normal expression domain of the respective Hox gene. In our system, the Hoxc-8 transgene is activated in a fashion that results in over-expression of Hoxc-8 in its normal domain of expression. The abnormalities in cartilage and bone formation in our Hoxc-8 transgenic animals thus support the interpretation that the transactivation of Hoxc-8 affects cell differentiation.

Several lines of evidence are consistent with this hypothesis: (1) Hoxc-8 is normally expressed in skeletogenic precursor cells and in chondrocytes at later stages of development (Bieberich et al., 1990; Yueh et al., 1998; and unpublished observations). (2) The Hoxc-8 promoter directs transgene expression specifically into skeletal structures and is excluded from developing skeletal muscle (Cormier-Regard, Yueh and Kappen, unpublished observations). (3) In the case of other Hox transgenes, they were typically controlled by a heterologous promoter resulting in ectopic expression. When the transgene was also overexpressed in its normal domain, patterning defects and the general variability of phenotypes in the conventional transgenic approach would have made detection of subtler changes in cell differentiation difficult (Pollock et al., 1995). (4) Furthermore, the temporal regulation of Hox gene expression progresses from an early pattern of expression throughout the somites and lateral mesoderm to a more restricted expression in skeletogenic precursor cells only (Kappen, 1998; Cormier-Regard and Kappen, unpublished observations). Interference in earlier processes (through ectopic expression) is likely to override effects at later developmental stages. (5) In our transgenic animals, Hoxc-8 expression is controlled by its own promoter. This promoter generally mimics the pattern of expression of the endogenous Hoxc-8 gene (Gardner et al., 1996; Cormier-Regard, Yueh, and Kappen, unpublished), except for possibly a short period around 10.5 days of development (Bieberich et al., 1990; Bradshaw et al., 1996; Belting et al., 1998). Thus, transactivation of Hoxc-8 in our binary system is more likely to create an over-expression situation rather than ectopic transgene activation.

To investigate the cellular basis of the cartilage defects in our Hoxc-8 transgenic mice, we isolated transgenic embryos at 16.5 and 17.5 days of development and produced histological sections (Yueh et al., 1998). Initial inspection of these sections indicated that there was an accumulation of cells in skeletal structures accompanied by a reduction in extracellular matrix. At the same time, there were fewer mature chondrocytes, which are characterized by a hypertrophic phenotype. In situ hybridization showed that there was elevated expression of collagen II, a marker for more immature chondrocytes. In addition, by staining with antibody against proliferating cell nuclear antigen (PCNA), we demonstrated that a significantly increased number of proliferative cells were present within skeletal structures but not in skeletal muscle or the spinal cord. Taken together, these data lead us to conclude that Hoxc-8 overexpression affects chondrocyte proliferation and/or differentiation (Yueh et al., 1998). More direct evidence for changes in cell proliferation comes from BrdU labeling in vivo in Hoxc-8 transgenic mice (Cormier-Regard and Kappen, unpublished observations). Taken together, our results suggest that Hoxc-8 is involved in regulating the proliferative capacity of skeletogenic cells and, in particular, acts at the level of the immature chondrocyte.

Comparative Transgenesis: Transactivation of Hoxc-8 in the Developing Nervous System.

The finding that Hoxc-8 regulates cell proliferation in cartilage led us to investigate whether Hoxc-8 could have a similar role during development of the nervous system. Because we had already generated the IE-Hoxc-8 TR strains, this experimental approach entailed only to produce TA strains in which the expression of VP16 was specifically targeted to cells in the nervous system. This condition was achieved by linking the VP16 transgene to the murine neurofilament light chain gene promoter (Yaworsky et al., 1997) or the second intron from the rat nestin gene (Zimmerman et al., 1994; Yaworsky and Kappen, 1999; and unpublished results). Crossing neuron-specific TA to the IE-LacZ TR lines confirmed that VP16 expression in these mice was specific to the developing nervous system (Yaworsky et al., manuscript in preparation).

Our expectation for these experiments was that neuron-specific transactivation of Hoxc-8 would extend its expression beyond its normal domain into more anterior neural structures such as the brain. If Hoxc-8 played a functional role in neural development, this situation would be expected to generate patterning defects. Alternatively, it could be possible that Hoxc-8 would affect the proliferation of neuronal precursor cells, which might similarly result in developmental defects. However we did not detect gross abnormalities or any appreciable changes in behavior or movement (Yaworsky et al., manuscript in preparation). These findings are in contrast to the phenotype in Hoxc-8-deficient animals, which exhibit abnormal fist clenching (LeMouellic et al., 1992) due to increased apoptosis of motorneuron subsets and abnormal axon projection patterns (Tiret et al., 1998). Although we cannot formally exclude that Hoxc-8 transgenic animals may develop subtle defects, it appears that ectopic and over-expression of Hoxc-8 does not critically alter neural tube development.

Combinatorial Transgenics I: Homeobox Genes In Neuronal Development.

To ascertain that the absence of a phenotype in these Hoxc-8 transgenic animals was not due to technical difficulties in transactivating a TR gene in neurons, we generated additional TR strains with an IE-Isl-1 transgene. Isl-1 had previously been shown to be important for motor-neuron differentiation (Pfaff et al., 1996), and we thus reasoned that its mis-expression might profoundly affect nervous-system development. This situation was indeed observed, in that embryos with transactivation of Isl-1 were severely retarded and lacked anterior neural structures. In addition, there were abnormalities in neural-tube folding and absence of closure (Yaworsky et al., manuscript in preparation). In fact, the double transgenic embryos died during gestation

Table 5.1. Phenotypes in Combinatorial Transgenics Using the VP16-Dependent Binary System

Transactivator	Activity (sites)	Transresponders		
		IE-Hoxc-8	IE-Hoxd-4	IE-Isl-1
Hoxc-8-VP16	early: posterior[a] late: cartilage[b]	Cartilage defect[b]	Cartilage defect[b]	posterior (growth) defect[c]
NF-VP16	neurons[d]	no phenotype?[e]	n. d.	n. d.
Nestin-VP16	neural precursors[e]	no phenotype?[e]	postnatal lethality[e]	embryonic lethality[e]

n. d. not done

[a] (Gardner et al., 1996)

[b] (Yueh et al., 1998)

[c] Muller, Yueh, Salbaum, and Kappen, unpublished observations

[d] (Yaworsky et al., 1997)

[e] (Yaworsky et al., manuscript in preparation)

(Table 5.1). Taken together, these results show that VP16-mediated transactivation is a feasible and valid approach to investigate the role of homeobox genes in neural development.

Combinatorial Transgenics II: Specificity of Hox Genes in Skeletal Development.

The VP16-dependent binary system also allowed us to address the specificity of homeobox transcription factors in cartilage development. An important question arising from our results was whether the over-expression of any homeodomain would produce the same consequences or if there was a need for a specific homeobox transcription factor. To approach this problem, we generated TR mouse strains for an IE-Hoxd-4 and IE-Isl-1 TR transgene. Our rationale was that by using the same transactivator, we could transactivate the other TR genes in exactly the same fashion, with regard to temporal and tissue specificity, as Hoxc-8. Activating a Hoxd-4 TR gene resulted in a phenotype morphologically similar to that found for the Hoxc-8 transgenic animals (Yueh et al., 1998). In contrast, the IE-Isl-1 transgenes did not produce cartilage abnormalities. This finding indicated that the role in chondrocyte proliferation/differentiation is likely specific to Hox-class homeodomain proteins. Thus, the possibility of combining different TR genes with the same TA strain allowed us to provide definitive evidence for a specific role of Hox genes in regulating cartilage development.

Combinatorial Transgenics III: Creation of Novel Phenotypes

In crossing the Hoxc-8-VP16 TA to IE-Isl-1 TR mice, we identified a novel unexpected phenotype. With two independent transactivators and 3 of 4 transresponders, we obtained progeny with shorter tails or caudal growth defects

(Muller, Yueh, Salbaum and Kappen, manuscript in preparation). Interestingly, the region most strongly affected by this defect reflects the early activity of the Hoxc-8 promoter (Table 5.1). In contrast, the cartilage defects in Hoxc-8 transgenic mice leave the posterior region unaffected and rather are consistent with the promoter's later tissue-specific activity. Conversely, transactivation of Isl-1 did not induce defects in the rib or axial cartilage that would be indicative of a functional role at later stages of development. These data provide strong evidence that different homeodomain transcription factors affect distinct stages and tissues in embryogenesis, even when expressed under control of the very same promoter. The excellent reproducibility and genetic stability in the VP16-dependent binary system now allows us to analyze, in detail, the pathogenesis of the phenotypes and their molecular basis.

4. SUMMARY

The applications of the VP16-dependent binary transgenic mouse system highlight its particular advantages: (1) The excellent specificity and reproducibility of VP16-mediated transactivation is particularly suited for investigations of developmental processes (Byrne et al., 1991). Although it may initially be appealing to contemplate systems in which transgene expression is inducible by small ligands, such approaches have been fraught with technical difficulties in delivering appropriate doses of ligand, teratogenic effects of ligand (Rossant and McMahon, 1999), and lack of temporal specificity due to the asynchronicity of embryos in a given litter (Byrne et al., 1991). In contrast, a defined promoter/enhancer to drive VP16 expression provides intricate temporal and tissue-specificity during development. (2) The initial efforts in establishing and maintaining independent TA and TR lines are clearly higher than with conventional technology. However, after a pair of TA and TR transgenic lines had been established (the Hoxc-8-VP16 TA and the IE-LacZ TR), the generation of additional transactivators did not require more transgenic strains than would have been generated in a conventional transgenic approach. Analogously, additional TR strains were created with the same effort that would have been expended conventionally. (3) In fact, the crosses of TA and TR mice could be performed directly with TR founder animals without the need for establishing lines. This approach allowed us to rapidly and simultaneously screen for functional TR loci as well as for the possible induction of phenotypic alterations. (4) The increased versatility in the binary system allows the combination of different TA and TR transgenes (Byrne et al., 1991). This flexibility enables truly controlled comparisons, either by transactivation of two different TR genes in the same tissue by the same transactivator or by transactivation of the very same TR gene in two distinct tissues, such as the Hoxc-8 locus in the nervous system and the skeletal system. (5) In studies of developmentally regulated molecules, the short time frame of developmental processes makes it important to avoid the variability inherent in conventional transgenic approaches as well as to circumvent potential lethality. As the best-characterized transactivation system, the VP16-dependent binary transgenic system provides a paradigm whose development is now clearly beyond the proof-of-concept stage. (6) Although it is

Table 5.2. Combinatorial Possibilities Using VP16-Mediated Transactivation

Transactivator	Transresponder
viral infection	IE-TR transgenic or knock-in
transgenic	IE-TR transgenic or knock-in
site-specific application of virus	IE-TR transgenic or knock-in
VP16-enhancer trap	IE-TR transgenic or knock-in
knock-in of VP16	IE-TR transgenic or knock-in
transgenic	site-specific application of virus
VP16-enhancer trap	site-specific application of virus
knock-in of VP16	site-specific application of virus

currently not (yet?) amenable to regulation by small ligands, VP16-mediated transactivation affords a variety of sophisticated approaches for the activation of transgenes. For example, a collection of VP16-TA mouse lines could be established from enhancer traps generated by random integration of microinjected VP16-transgenes or random mutagenesis in embryonic stem (ES) cells (Gossler et al., 1989). In addition, VP16 can be used to disrupt an endogenous gene. This approach would not only lead to targeted mutagenesis of the gene locus but at the same time allow transactivation of a reporter or of functional transgenes. In this way, any transgene could be activated in a very precise manner (regulated through the endogenous promoter/enhancers of the targeted locus), effectively replacing one gene product with another in a specific set of cells or tissues. Such precise regulation is seldom possible with conventional transgenic approaches. Furthermore, providing VP16 through stereotactic injection into TR mice enables a controlled delivery of transactivation to specific tissues at defined times. Analogously, the transgene to be transactivated may be supplied in form of viral infection or other means of gene transfer in vivo. Table 5.2 lists possible applications of VP16-mdiated transactivation in the context of transgenic or mutant mice and viral infection. More complicated scenarios are also conceivable, such as placing the VP16 transactivation system on genetic backgrounds that selectively limit, or supply in controlled or regulatable manner, the cooperating factors needed for VP16 transactivation. Finally, a variety of combinations of transactivation and recombination systems can be imagined, limited in complexity and practicability only by principles of mouse reproduction and genetics.

Methods and Technical Information

Transgenes and Mice. Transgenic mice were generated by standard methods (Hogan et al., 1986) and have been previously described (Gardner et al., 1996; Yaworsky et al., 1997; Rundle et al., 1998; Yueh et al., 1998).

Genotyping for Transgenes. Transgenic founder animals were identified by PCR on genomic DNA obtained from tail samples as described previously (Yaworsky et al., 1997). Southern blot analysis confirmed the integrity of the respective transgenes (Gardner et al., 1996; Yaworsky et al., 1997; and unpublished data). After establishment of the first-generation offspring from founder animals, further genotyping was routinely performed by PCR. To distinguish progeny homozygous and hemizygous for a given transgene locus, we performed semi-quantitative PCR (Rundle et al., 1998; Yueh et al., 1998; and unpublished results). Briefly, the DNA concentration for each tail sample was determined spectroscopically. PCR reactions were performed for a specific number of cycles that had been empirically defined to represent the reaction's linear range for each particular primer pair (Yueh et al., 1998). The intensity of EtBr-staining of PCR products electrophoresed on standard 1% agarose gels was measured by digitizing Polaroid photographs and using the histogram function of Adobe Photoshop 3. The arbitrary units of density (histogram density units, HDUs) for the same window size per band were calculated by using the following formula (where [DNA] is the DNA concentration for each sample):

$$\frac{HDU}{[DNA]} \text{ divided by the } \frac{HDU}{[DNA]} \text{ for a known hemizygote positive-control sample.}$$

The results were close to either 1 or 2, indicating hemi- or homozygosity, respectively. This assessment was confirmed, for each independent strain, by breeding to wild-type mice; the transgene transmission ratio from mice typed as hemizygotes was 50%, from those typed as homozygotes, 100% (Yueh, Smith and Kappen, unpublished data).

Exclusion of Detrimental Transgene Insertion Sites. To exclude the possibility that the transgene insertion sites would have detrimental effects on transgene expression or embryo development, we crossed mice hemizygous for their transgene locus to homozygosity. Our expectation was that potential disruption of critical genomic regions by transgene insertion would be revealed in abnormal inheritance or development defects. We have not detected any obvious insertional mutants amongst any of the reported transgenic strains yet. These observations indicate that none of the insertion sites disrupted an essential gene. Secondly, increased VP16 expression due to increased transgene doses evidently had no detrimental effects. Finally, if any of the tested TR transgenes was expressed at basal levels, they did not cause any obvious or detrimental developmental abnormalities.

Expression of the VP16 Transactivator. Transactivator transgene expression was assayed by three methods: (1) immunohistochemistry using a monoclonal antibody (Lyons and Chambon, 1995; generously provided by P. Chambon, Strasbourg, France); (2) RT-PCR with VP16-specific primers; and (3) by way of transactivation of β-galactosidase expression from an IE-LacZ TR gene. We were unable to detect VP16 at the protein level (unpublished observation) whereas RT-PCR and functional transactivation proved to be effective measures of detecting VP16 expression (Rundle et al., 1998). Northern blot analysis may be suitable to detect VP16 mRNA in adult transgenic tissues; however, because our studies focused on embryonic stages, the low yields of RNA from embryonic tissues precluded the use of this method. Taken together, our results suggest that VP16 expression may be below the level of detection afforded by immunohistochemistry but is clearly measurable by the more sensitive PCR technique and by biological (i.e., transactivation) activity. Notably, we observed that VP16 transactivator expression was only moderately persistent. For example, neurons in the regions of the embryonic spinal cord that displayed LacZ staining at 13.5 days exhibited little or no evidence of transactivation by 14.5 days (Yaworsky et al., 1997). Similarly, VP16 activity from the Hoxc-8-VP16 transactivator that was present in muscle precursor cells in the posterior region at 12.5 days was no longer detectable in 14.5-day-old embryos, clearly indicating elimination of VP16 from these cells within 2 days or less. Thus, VP16 expression in transgenic mice appears to be dependent on continuous activity of the respective promoter in TA mice.

Quantitation of Transactivator Activity. We measured the functional activation activity of VP16 in vivo by chemiluminescence assay for β-galactosidase activity. The relative levels of enzyme activity in whole embryo extracts were compared for mice transgenic for transactivator only, transresponder only, and double transgenic mice (Rundle et al., 1998). There were significant increases of enzyme activity in the double transgenic embryos over the background chemiluminescence in TA or TR only samples. The relative levels of activity ranged from 1.6-fold to 25.6-fold (Rundle et al., 1998), indicating the "strength" of transactivation for a given TA line. We also measured the relative levels of enzyme activity in different regions of embryos that we isolated by microdissection. The IE-LacZ transgene was expected to be predominantly activated in the posterior region (Bieberich et al., 1990; Gardner et al., 1996). This measurement allowed us to show that despite small amounts of β-galactosidase activity in anterior regions (in which no transactivation would be expected), there was specific elevation of enzyme activity by up to 26.25-fold by VP16-mediated transactivation (Rundle et al., 1998). Although these measurements were influenced by the precision of dissection of the embryonic tissues, there was consistent reproducibility. We found as much as 15-fold difference in transactivation ability of different TA lines (Rundle et al., 1998).

Quantitation of Transresponder Gene Expression. To detect TR gene expression, we mated proven transactivators to TR transgenic animals and isolated the posterior regions from 10.5-day embryos of such a cross (Rundle et al., 1998). After genotyping the respective yolk sac DNAs, we assayed for the presence of TR message by RT-PCR. In this way, we detected activation of the TR transgene in about 70% of 20 independent TR strains tested (Rundle et al., 1998). To get a quantitative estimate of the activation of TR gene expression, we devised a RT-PCR assay in which the expression of Hoxc-8 TR genes was measured relative to the level of expression from the endogenous Hoxc-8 gene in the same region of the embryo. The activation of TR gene expression ranged from 1.8- to 6.7-fold (Rundle et al., 1998). This assay is not particularly precise, especially when the transactivator does not fully reproduce the levels of endogenous gene expression in all cells and tissues, or because temporal activity of a given transactivator may differ from the endogenous locus (as in the case of the nestin gene enhancer; Yaworsky and Kappen, 1999). Furthermore, we have found evidence that the activity level of TA transgenes may be temporally modulated (presumably by the integration site), even when spatial specificity was undisturbed (unpublished observations). We therefore confirmed for all VP16 transactivators, their activity at multiple points in embryogenesis by crossing the TA's to the IE-LacZ reporter TR strains (Gardner et al., 1996; Yaworsky et al., 1997). As discussed previously, however, the quantitative aspects of transactivation at different embryonic stages in the VP16 system remain to be elucidated.

Gene Dosage Control in VP16-Mediated Transactivation. Cases in which animals hemizygous for both the TA and TR locus were viable allowed us to increase transgene dosage by superimposition of transgene loci (Rundle et al., 1998; Yueh et al., 1998; manuscript in preparation). As indicated by the measurements of β-galactosidase enzyme activity, the gene dosage of the VP16 TA transgene appeared to be more effective in increasing TR transactivation than homozygosity at the TR transgene locus (see also effects of Hoxc-8 transactivation on cartilage differentiation; Yueh et al., 1998). In addition, we were unable to overcome the lack of obvious phenotypes in mice with transactivation of Hoxc-8 in the nervous system by increases in TR transgene dosage (Yaworsky et al., manuscript in preparation). Although these results have not been systematically confirmed for all possible TA/TR combinations, they suggest that the amount of target TR is less limiting than the available levels of VP16 TA expression. This situation allows us to activate the same TR locus and create the analogy of an "allelic" series by using transactivators of different "strength" and specific combinations of hemi- and homozygosity at the TA and/or TR transgene loci (see Fig. 5.3). Because the expected frequency of progeny that are homozygous for one or more of the transgene loci is ~31% combined from a cross of two double hemizygotes (TA/+ TR/+ x TA/+ TR/+, see Fig. 5.3), we also set up crosses of a double hemizygote with a mouse that was homozygous at either the TA or TR locus (TA/+ TR/+ x TA/TA +/+ or TA/+ TR/+ x +/+ TR/TR). In these cases, all progeny will be homozygous for one of the loci and either inherit the second locus in hemizygous form or not at all (outcomes: TA/TA TR/+ and TA/TA +/+ from the first cross, and TA/+ TR/TR and +/+ TR/TR from the second possibility). This combination of genotypes raises the expected frequency of desired genotypes to 50% and at the same time generates control animals that lack transactivation altogether.

ACKNOWLEDGMENTS

I thank Drs. Stephania Cormier-Regard and Y. Gloria Yueh for providing data for Figure 5.2. I am also grateful to Drs. Guerard W. Byrne, Pierre Chambon, Mark Featherstone, Andrew P. McMahon and J. Michael Salbaum for gifts of reagents; to Drs. David P. Gardner, MiMi P. Macias, Frank H. Ruddle, Charles H. Rundle, Manuel F. Utset, J. Michael Salbaum, Paul J. Yaworsky, and Y. Gloria Yueh for their contributions to the development of the VP16-dependent transgenic system and discussions; to Steve Farmer, Jean Kloss, Kristin Smith, and Teresa Tinder for expert technical assistance and to the Mayo Clinic Scottsdale Core Facilities for Histology (M. Anita Jennings), Transgenic Animals

(Stephanie Munger, Suresh Savarirayan, and Dr. Sergej Ochkur), and Visual Communications (Julie Jensen and Marv Ruona). Carol Williams helped with the preparation of the manuscript. Work in my laboratory on projects described here was funded by the American Parkinson's Disease Association, the National Parkinson's Foundation, the Arizona Disease Control Research Commission, the Arthritis Foundation, Mayo Foundation for Medical Education and Research, and NIH grants NIAMS R03-AR44945 and NICHD R01-HD37804.

REFERENCES

Balling, R, Mutter G, Gruss P, Kessel M (1989): Craniofacial abnormalities induced by ectopic expression of the homeobox gene Hox-1.1 in transgenic mice. Cell 58:337–347.

Belting HG, Shashikant CS, Ruddle FH (1998): Multiple phases of expression and regulation of mouse Hoxc-8 during early embryogenesis. J Exp Zool 282:196–222.

Bieberich CJ, Utset MF, Awgulewitsch A, Ruddle FH (1990): Evidence for positive and negative regulation of the Hox-3.1 gene. PNAS USA 87:8462–8466.

Boissiere SL, HughesT, O'Hare P. (1999): HCF-dependent nuclear import of VP16. EMBO J 18:480–489.

Bradshaw MS, Shashikant CS, Belting HG, Bollekens JA, Ruddle FH (1996): A long-range regulatory element of Hoxc8 identified by using the pClasper vector. PNAS USP NAS US 93:2426–2430.

Byrne GW, Kappen C, Schughart K, Utset M, Bogarad L, Ruddle FH (1991): Analysis of controller genes using the transgenic mouse system. In First N, Haseltine F (eds) "Transgenic Animals" Biotechnology 16:135–152.

Byrne GW, Ruddle FH (1989): Multiplex gene regulation: a two-tiered approach to transgene regulation in transgenic mice. PNAS USA 86: 5473–5477.

Capecchi MR (1996): Function of Homeobox Genes in Skeletal Development. Ann NY Acad 785:34–37.

Charite J, deGraaff W, Shen SB, Deschamps J (1994): Ectopic expression of hoxb-8 causes duplication of the zpa in the forelimb and homeotic transformation of axial structures. Cell 78:589–601.

Chen F, Capecchi MR (1997): Targeted mutations in hoxa-9 and hoxb-9 reveal synergistic interactions. Dev Biol 181:186–196.

Duboule D (1995): Vertebrate Hox-genes and proliferation: an alternative pathway to homeosis? Cur Op Gen 5:525–528.

Freiman RN, Herr W (1997): Viral mimicry: common mode of association with HCF by VP16 and the cellular protein LZIP. Gene Dev 11:3122–3127.

Gardner DP, Byrne GW, Ruddle FH, Kappen C (1996): Spatial and temporal regulation of a LacZ reporter transgene in a binary transgenic mouse system. Transgen Re 5:37–48.

Gardner DP, Kappen C (1998): Two novel transgene insertions on the mouse Y chromosome. Transgenics 2:221–224.

Gerster T, Roeder RG (1988): A herpesvirus trans-activating protein interacts with transcription factor OTF-1 and other cellular proteins. PNAS USA 85:6347–6351.

Gossler A, Joyner AL, Rossant J, Skarnes WC (1989): Mouse embryonic stem cells and reporter constructs to detect developmentally regulated genes. Science 244:463–465.

Gunning P, Leavitt J, Muscat G, Ng S-Y, Kedes L (1987): A human β-actin expression vector system directs high-level accumulation of antisense transcripts. PNAS USA 84:4831–4835.

Hanahan D (1989): Transgenic mice as probes into complex systems. Science 246:1265–1275.

He Z, Brinton BT, Greenblatt J, Hassell AJ, Ingles CJ (1993): The transactivator proteins VP16 and GAL4 bind replication factor A. Cell 73:1223–1232.

Hengartner CJ, Thompson CM, Zhang J, Chao DM, Liao SM, Koleske AJ, et al. (1995): Association of an activator with an RNA polymerase II holoenzyme. Gene Dev 9:897–910.

Hogan B, Constantini F, Lacy E. Manipulating the Mouse Embryo: A Laboratory Manual. Cold Spring Harbor Laboratory: Cold Spring Harbor, N.Y, 1986.

Horan GSB, Ramirez-Solis R, Featherstone MS, Wolgemuth DJ, Bradley A, Behringer RR (1995): Compound mutants for the paralogous hoxa4, hoxb-4, and hoxd-4 genes show more complete homeotic transformations and a dose-dependent increase in the number of vertebrae transformed. Gene Dev 9:1667–1677.

Ingles CJ, Shales M, Cress WD, Triezenberg SJ, Greenblatt J (1991): Reduced binding of TFIID to transcriptionally compromised mutants of VP16. Nature 351:588–590.

Jegalian BG, DeRobertis EM (1992): Homeotic transformations in the mouse induced by overexpression of a human Hox3.3 transgene. Cell 71:901–910.

Kappen C. Early and late functions of homeobox genes in the development of the axial skeleton. In Skeletal Growth and Development: Clinical Issues and Basic Science Advances; Buckwalter JA, Ehrlich MG, Sandell LJ, Trippel SB, Eds.; American Academy of Orthopedic Surgeons: Rosemont, IL, 1998; pp 147–162.

Kaur S, Singh G, Stock JL, Schreiner CM, Kier AB, Yager KL, et al. (1992): Dominant mutation of the murine Hox-2.2 gene results in developmental abnormalities. J Exp Zool 264:323–336.

Kobayashi N, Boyer TG, Berk AJ (1995): A class of activation domains interacts directly with TFIIA and stimulates TFIIA-TFIID-promoter complex assembly. Mol Cell Biol 15:6465–6473.

Koyasu S, Hussey RE, Clayton LK, Lerner A, Pedersen R, Delany-Heiken P, et al. (1994): Targeted disruption within the CD3 zeta/eta/phi/Oct-1 locus in mouse. EMBO J 13:784–797.

Kristie TM, LeBowitz JH, Sharp PA (1989): The octamer-binding proteins form multi-protein DNA complexes with HSV aTIF regulatory protein. EMBO J 8:4229–4238.

LeMouellic H, Lallemand Y, Brulet P (1992): Homeosis in the mouse induced by a null mutation in the Hox-3.1 gene. Cell 69:251–264.

Lin YS, Green MR (1991): Mechanism of action of an acidic transcriptional activator in vitro. Cell 64:971–981.

Lu R, Yang P, Padmakumar S, Misra V (1998): The herpesvirus transactivator VP16 mimics a human basic domain leucine zipper protein, luman, in its interaction with HCF. J Virology 72:6292–6297.

Lyons JG, Chambon P (1995): Direct activation and anti-repression functions of GAL4-VP16 use distinct molecular mechanisms. Biochem J 312:899–905.

McGinnis W, R Krumlauf. (1992): Homeobox genes and axial patterning. Cell 68: 283–302.

McKnight JL, Kristie TM, Roizman B (1987): Binding of the virion protein mediating alpha gene induction in herpes simplex virus 1-infected cells to its cis site requires cellular proteins. PNAS USA 84:7061–7065.

McLain K, Schreiner C, Yager KL, Stock JL, Potter SS (1992): Ectopic expression of Hox-2.3 induces craniofacial and skeletal malformations in transgenic mice. Mech Devel 39:3–16.

Mitchell W (1995): Neurons differentially control expression of a herpes simplex virus type 1 immediate-early promoter in transgenic mice. J Virology 69:7942–7950.

Ohno H, Aoe T, Taki S, Kitamura D, Ishida Y, Rajewsky K, Saito T (1993): Developmental and functional impairment of T cells in mice lacking CD3 zeta chains. EMBO J 12:4357–4366.

Pfaff SL, Mendelsohn M, Stewart CL, Edlund T, Jessell TM (1996): Requirement for LIM homeobox gene Isl-1 in motor neuron generation reveals a motor neuron-dependent step in interneuron differentiation. Cell 84:309–320.

Pollock RA, Jay G, Bieberich CJ (1992): Altering the boundaries of Hox3.1 expression: evidence for antipodal gene regulation. Cell 71:911–923.

Pollock RA, Sreenath T, Ngo L, Bieberich CJ (1995): Gain of function mutations for paralogous Hox genes: Implications for the evolution of Hox gene function. PNAS USA 92:4492–4496.

Preston CM, Frame MC, Campbell ME (1988): A complex formed between cell components and an HSV structural polypeptide binds to a viral immediate early gene regulatory DNA sequence. Cell 52:425–434.

Rijli FM, Chambon P (1997): Genetic interactions of Hox genes in limb development: learning from compound mutants. Cur Op Gen 7:481–487.

Rossant J, McMahon A (1999): "Cre"-ating mouse mutants—a meeting review on conditional mouse genetics. Gene Dev 13:142–145.

Rundle CH, Macias MP, Gardner DP, Yueh YG, Kappen C (1998): Transactivation of Hox Gene Expression in a VP16-Dependent Binary Transgenic Mouse System. Biochem Biophys Acta 1398:164–178.

Stern S, Tanaka M, Herr W (1989): The Oct-1 homeodomain directs formation of a multiprotein-DNA complex with the HSV transactivator VP 16. Nature 341:624–630.

Stringer KF, Ingles CJ, Greenblatt J (1990): Direct and selective binding of an acidic transcriptional activation domain to the TATA-box factor TFIID. Nature 345:783–786.

Suzuki N, Peter W, Ciesiolka T, Gruss P, Schöler HR (1993): Mouse Oct-1 contains a composite homeodomain of human Oct-1 and Oct-2. Nucl Acid R 21:245–252.

Tiret L, LeMouellic H, Maury M, Brulet P (1998): Increased apoptosis of motoneurons and altered somatotopic maps in the brachial spinal cord of Hoxc-8-deficient mice. Develop 125:279–291.

Westphal H, Gruss P (1989): Molecular genetics of development studied in the transgenic mouse. Ann Rev Cell Biol 5:181–196.

Wilson AC, LaMarco K, Peterson MG, Herr W (1993): The VP16 accessory protein HCF is a family of polypeptides processed from a large precursor protein. Cell 74:115–125.

Wold MS (1997): Replication protein A: a heterotrimeric, single-stranded DNA-binding protein required for eukaryotic DNA metabolism. Ann R Bioch 66:61–92.

Yaworsky PJ, Gardner DP, Kappen C (1997): Transgenic analyses reveal neuron and muscle specific elements in the murine neurofilament light chain gene promoter. J Biol Chem 272:25112–25120.

Yaworsky PJ, Kappen C (1999): Heterogeneity of neural progenitor cells revealed by enhancers in the nestin gene. Dev Biol 205:309–321.

Yaworsky PJ, Salbaum JM, Kappen C. Transgenic studies on the role of homeobox genes in nervous system development. manuscript in preparation.

Yueh YG, Gardner DP, Kappen C (1998): Evidence for regulation of cartilage differentiation by the homeobox gene Hoxc-8. PNAS USA 95:9956–9961.

Yueh YG, Yaworsky PJ, Kappen C. The Herpes Simplex Virus transcriptional transactivator VP16 is detrimental to preimplantation development in the mouse. submitted for publication.

Zhang MB, Kim HJ, Marshall H, Gendronmaguire M, Lucas DA, Baron A, et al. (1994): Ectopic hoxa-1 induces rhombomere transformation in mouse hindbrain. Develop 120:2431–2442.

Zimmerman L, Lendahl U, Cunningham M, Mckay R, Parr B, Gavin B, Mann J, et al. (1994): Independent regulatory elements in the nestin gene direct transgene expression to neural stem cells or muscle precursors. Neuron 12:11–24.

CHAPTER 6

GENE-TARGETING STRATEGIES IN THE ANALYSIS OF ALCOHOL AND ANESTHETIC MECHANISMS

GREGG E. HOMANICS, PH.D.

I. INTRODUCTION

The molecular mechanisms responsible for alcohol and anesthetic action have evaded elucidation despite intense investigation. More than 150 years ago, diethyl ether was demonstrated to be an effective anesthetic capable of inducing amnesia and immobility and blocking pain. Despite the fact that the discovery of anesthesia is perhaps one of the greatest medical advances of all time, it remains unknown how anesthetics induce their clinically important effects in a readily reversible manner. Alcohol has been consumed for thousands of years. Today, alcohol is society's most abused drug. Studies estimate that the economic cost of alcohol abuse in 1995 in the United States alone was more than $166 billion.[1] To those directly affected by alcohol abuse, the personal toll is incalculable. Despite the ubiquity of alcohol use and abuse, it is unclear how this very simple molecule exerts its diverse array of physiological and behavioral effects. Understanding the molecular mechanism of alcohol within the body, and especially in the nervous system, will ultimately lead to new therapeutic interventions to combat this devastating disease.

More than a century of research into the mechanism of action of these neuroactive substances suggests that common neuronal substrates are responsible for alcohol and anesthetic action. Earlier dogma held that alcohol and anesthetics disrupted higher-order cognitive function in a non-specific manner by interacting with membrane lipids, that is, a lipid-based theory of action. More recently, most researchers in the field now believe that alcohol and anesthetics directly interact in a very specific manner with proteins to exert their effects,

[1] Dollar figures are from "The Economic Costs of Alcohol and Drug Abuse in the United States" by the National Institute on Drug Abuse and the National Institute on Alcohol Abuse and Alcoholism. This is available online at: http://www.nida.nih.gov/EconomicCosts/Index.html.

Genetic Manipulation of Receptor Expression and Function,
Edited by Domenico Accili.
ISBN 0-471-35057-5   Copyright © 2000 Wiley-Liss, Inc.

Table 6.1. Potential Targets of Alcohol and Anesthetic Action

Acetylcholine receptors
Adenylate cyclase
Alpha adrenergic receptors
Dopamine receptors
GABA type A receptors
Glycine receptors
Glutamate receptors
Neuropeptide Y receptors
Nitric oxide
Opioid receptors
Protein kinases
Serotonin (5HT) receptors
Others?

that is, a protein-based theory (Franks and Lieb, 1994). Work from the past decade indicates that the most plausible protein targets are the ligand-gated ion channels that function to control synaptic transmission (Table 6.1). Much of the previous effort has utilized in vitro approaches and simple model systems. Although these studies have been very informative, it must be recognized that clinically relevant responses to anesthetics and alcohol occur at the organism level and are characterized by a wide range of behavioral end points. At low doses, these drugs induce ataxia, increased locomotion and anxiolysis; at higher doses, amnesia and the inability to feel pain; at extremely high doses, coma and ultimately death. Thus, the ultimate level of alcohol and anesthetic responses must be analyzed in a system that is capable of demonstrating complex behavioral responses, that is, the intact animal.

The wide variety of behavioral perturbations induced by alcohol and anesthetics is unlikely to be due to effects on a single protein but more likely reflect actions on multiple protein targets. The fact that multiple targets are involved has made it extremely difficult to tease out the contribution of individual proteins to the effects of these drugs in vivo. That is, until recently when the gene targeting and transgenic approaches have been applied to this problem.

Gene targeted animals are those organisms that have had an endogenous gene modified by the process of homologous recombination. Transgenic animals are those organisms that have had exogenous genes added to their genetic repertoire. Together, gene-targeted and transgenic animals are generically referred to as "genetically engineered organisms." To date, mice are the primary genetically modified organism that has advanced our understanding of the mechanism of action of alcohol and anesthetics.

This chapter is not intended to be an exhaustive review of all genetically engineered mice that have been created and utilized to investigate the mechanism of action of alcohol and anesthetics. Instead, this chapter will provide an overview of how these mice have helped shape the current state of knowledge of how these drugs exert their effects. Several examples that illustrate the power and usefulness, as well as the limitations of the genetic engineering approach, will be highlighted. Table 6.2 lists many of the strains of genetically modified

Table 6.2. Genetically Altered Mice That Have Been Used in Alcohol and Anesthesia Research

Mouse	Relevant Phenotype	Selected References
GABA$_A$ receptors		
alpha 6 subunit knockout	normal response to ethanol; more sensitive to diazepam	(Homanics et al., 1997b; Homanics et al., 1998a; Korpi et al., 1999)
beta 3 subunit knockout	reduced sensitivity to some anesthetics for some endpoints	(Quinlan et al., 1998)
gamma 2 subunit knockout	insensitive to benzodiazepines	(Gunther et al., 1995)
gamma 2 long subunit knockout	normal response to ethanol	(Homanics et al., 1999)
Serotonin receptors		
5-HT$_{1B}$ knockout	increased ethanol consumption; decreased sensitivity and tolerance to ethanol	(Crabbe et al., 1996)
5-HT$_3$ transgenic	reduced ethanol consumption	(Engel et al., 1998)
Opiate receptors		
mu knockout	insensitive to morphine	(Kieffer, 1999; Loh et al., 1998; Matthes et al., 1996; Sora et al., 1997b; Tian et al., 1997)
kappa knockout	insensitive to kappa specific agonists; attenuates morphine withdrawal	(Kieffer, 1999; Simonin et al., 1998)
Alpha adrenergic receptors		
alpha 2a knockin	insensitive to dexmetomidine	(Lakhlani et al., 1997)
alpha 2c knockout	decreased hypothermic effect of dexmetomidine	(Sallinen et al., 1997)
alpha 2c transgenic	increased hypothermic effect of dexmetomidine	(Sallinen et al., 1997)
Dopamine receptors		
D2 knockout	abolished rewarding effect of morphine	(Maldonado et al., 1997)
D4 knockout	supersensitive to ethanol, cocaine, and methamphetamine	(Rubinstein et al., 1997)
Kinases		
fyn tyrosine kinase knockout	increased sensitivity to ethanol	(Miyakawa et al., 1997)
protein kinase C gamma knockout	decreased sensitivity to ethanol	(Harris et al., 1995)
Others		
neuronal nitric oxide synthase knockout	normal anesthetic sensitivity	(Crosby et al., 1995; Ichinose et al., 1995)
neuropeptide Y knockout	increased ethanol consumption; decreased ethanol sensitivity	(Thiele et al., 1998)
neuropeptide Y transgenic	decreased ethanol consumption; increased ethanol sensitivity	(Thiele et al., 1998)

mice that have been applied to alcohol and anesthesia research. For a more comprehensive discussion of these rodents, the reader is referred to a recent review (Homanics et al., 1998b).

2. ALCOHOL

Clearly alcoholism is a disease that is in part under genetic control. Alcoholism is no longer considered a purely psychological disease. It is well-recognized that there are underlying genetic factors that predispose individuals to alcohol abuse. Alcoholism has long been known to run in families; first-degree relatives of alcoholics are seven times more likely to develop alcoholism than controls (Merikangas, 1990). Monozygotic (identical) twins have a greater concordance for alcoholism than do dizygotic (fraternal) twins (Cadoret, 1990). Adoption studies indicate that children of alcoholics have an ~2.5-fold increased risk for alcoholism, irrespective of the home environment in which they were raised (Merikangas, 1990). Animal model studies also clearly demonstrate a genetic component to all facets of alcohol use (Crabbe et al., 1994). Thus, there is no longer any doubt that genes play a significant role in alcoholism. The overriding question at present is which of the 70,000–100,000 genes in the genome modulate the various effects of alcohol.

With the gene-targeting approach, one can analyze the influence of single known genetic perturbations on alcohol-induced endpoints in intact animals. Despite the fact that this approach has only recently been used to investigate the molecular site of action of alcohol, incredibly exciting insight has already been gained. Several examples are presented below.

Serotonin (5-hydroxytryptamine; frequently abbreviated as 5-HT) is a neuromodulator that affects the release of neurotransmitters such as dopamine, acetylcholine, and γ-aminobutyric acid (GABA). Abnormal functioning of the serotonin neurotransmitter system has been implicated in alcoholism as well as in other neurological disorders, for example, anxiety, aggressive tendencies, and suicide. With a single exception, membrane-spanning receptors for serotonin (14 have been identified to date) are all linked to G-proteins, and thus function through second messengers. The exception is the 5-HT_3 receptor, which is a ligand-gated sodium channel.

Mice lacking the 5-HT_{1B} receptor have been created by gene targeting (Saudou et al., 1994). These mice display a wide range of pharmacological and behavioral alterations, including failure to respond to the 5-HT_1 agonist RU24969 and enhanced aggression. These knockouts have recently been scrutinized for alterations in alcohol-related behaviors (Crabbe et al., 1996). Although most mice are very reluctant to voluntarily consume alcohol, 5-HT_{1B} knockout mice readily consume significantly more alcohol (including solutions containing up to 20% ethanol) than do controls. Consumption of food, water, and sweet and bitter tastants was unaffected. This finding indicates that inactivation of the 5-HT_{1B} receptor had a very specific effect on ethanol consumption. Knockout of the 5-HT_{1B} receptor also altered the acute ataxic response to ethanol (knockout mice were more sensitive to ethanol) and the development of tolerance following repeated exposure (knockouts developed tolerance more slowly than did control mice). In contrast, the withdrawal reaction following either acute or chronic exposure to ethanol was unaffected by the genetic perturbation. Together these results indicate that elimination of the 5-HT$_{1B}$ receptor leads to reduced sensitivity to ethanol on some behavioral end points such as consumption, motor coordination, and tolerance, but this genetic change has no effect on dependence. This observation of response-specific changes is important in that it conclusively establishes that the wide range of ethanol-induced behaviors is under separate, but perhaps partially overlapping, genetic control. Thus, it is prudent to examine as many different behavioral end points as possible when testing genetically modified animals for response to ethanol so that interesting and exciting phenotypes are not overlooked.

Involvement of the ligand-gated 5-HT$_3$ receptor in the reward pathway that modulates behavioral response to ethanol has been suggested from pharmacologic studies in animals and humans. Transgenic mice overexpressing this receptor specifically in the forebrain have recently been produced and tested for propensity to consume ethanol (Engel et al., 1998). It is interesting to note that the 5-HT$_3$ transgenic mice consumed almost 50% less ethanol than did wild-type control mice. This reduction in ethanol consumption was specific in that food intake and total fluid intake did not differ between genotypes. These results suggest that over-expression of the 5-HT$_3$ receptor in the forebrain diminished the rewarding effects of ethanol. Presumably, additional studies investigating the multitude of ethanol-responsive behavioral end points are in progress with these interesting genetically engineered animals.

Neuropeptide Y (NPY) is a 36-amino-acid peptide that functions as a neuromodulator by signaling through receptors that are coupled to G-proteins, which ultimately regulate intracellular levels of cAMP. From genetic linkage studies of a selectively bred rat line that models alcoholism, it has been suggested that NPY may be involved in modulating alcohol-related behavioral responses (Carr et al., 1998).

To test directly the involvement of NPY in alcohol-related behaviors, genetically engineered mice that either lacked NPY (knockout mice) or over-expressed NPY (transgenics) were studied (Thiele et al., 1998). In tests of ethanol consumption, NPY knockouts consistently consumed significantly more ethanol than did wild-type animals, whereas the NPY transgenics consumed significantly less ethanol compared with controls. Neither the knockouts or transgenics differed from controls in preference for sweet (sucrose) or bitter (quinine) tastants, average food intake, or average total fluid consumption (ethanol + water). These results indicate that the observed changes in ethanol consumption are specific for ethanol and the changes are not calorie-driven. The sedative/hypnotic effects of ethanol were also investigated in these genetically engineered rodents. NPY knockouts were less sensitive than controls in a sleep-time assay, that is, sleep-time in response to a single acute ethanol injection was significantly reduced in knockouts relative to controls. Conversely, sleep-time response of the NPY over-expressing mice was dramatically extended compared with wild-type controls. These differences in the sedative/hypnotic effects of ethanol were not due to alterations in clearance or metabolism of ethanol in the mice. Together, these exciting results from mice with specific alterations in the NPY gene indicate that ethanol consumption and resistance are specifically modulated in an inverse manner by NPY levels in the central nervous system.

Whereas some gene knockout mice, such as those just described, have yielded dramatic alcohol-related phenotypes, other knockouts have yielded stunning results by refuting the role of putative components of ethanol's molecular mechanism of action. Such is the case for the knockout of the long splice variant of the gamma 2 subunit (gamma 2 long) of the GABA type A receptor (Homanics et al., 1999).

GABA is the primary inhibitory neurotransmitter in the mammalian central nervous system (for review, see: Sieghart, 1995). The inhibitory effects of GABA are mediated by the GABA type A receptor. These receptors are het-

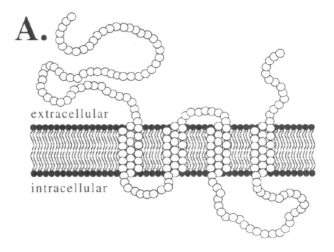

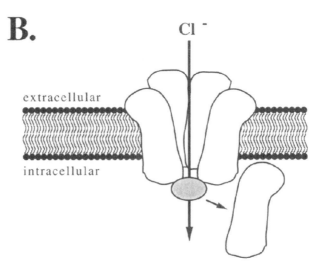

Figure 6.1. GABA type A receptor structure. A. Diagram of a single receptor subunit polypeptide (chain of open circles) illustrating a large amino terminal extracellular domain and four transmembrane spanning domains that pass through the lipid bilayer. B. Diagram of a heteropentameric receptor illustrating the central pore that forms the chloride channel.

eropentameric and are constructed from at least 17 different subunits that can be grouped into families based on homology (alpha 1-6, beta 1-4, gamma 1-3, delta, epsilon, and rho 1-2). The various subunits are each expressed in unique temporal and tissue-specific patterns in the nervous system. Each subunit polypeptide is a membrane-spanning protein that is thought to traverse the neuronal membrane four times (Figure 6.1A). The subunits of the heteropentamer are thought to assemble into a donut structure with a central pore (Fig. 6.1B). Binding of GABA to the receptor causes chloride ions to flow into the cell through the pore, hyperpolarizing the cell, ultimately resulting in neuronal inhibition. The stoichiometry of physiologically important GABA type A receptor isoforms is just beginning to be elucidated (McKernan and Whiting, 1996). It appears that the majority of receptors are composed of two alpha, two beta, and one gamma subunit. These receptors are prime candidates for the molecular site of action of alcohol (and anesthetics). In vitro, it is clear that at physiologically relevant concentrations, alcohol can potentiate the effects of GABA at some GABA type A receptors. However, in vivo at the whole animal level, the contribution of alcohol actions at the GABA type A receptor to relevant behavioral end points is largely unknown as is the importance of the numerous individual subunit polypeptides. The gene knockout approach allows one to genetically dissect the contributions of the individual components of this extraordinarily complex receptor system to cellular and whole animal responses to alcohol.

Although highly controversial, it has been reported that specifically gamma 2 long—but not the short splice variant of the gamma 2 subunit of the GABA type A receptor—was absolutely critical for ethanol potentiation of the effects of GABA at the receptor (Wafford et al., 1991). Furthermore, it was demonstrated that protein kinase C mediated phosphorylation of the gamma 2 long subunit was the critical determinant that regulated sensitivity of the GABA type A receptor to ethanol (Wafford and Whiting, 1992). These experiments were all conducted with in vitro expression systems. Thus, the in vivo whole animal relevance of the gamma 2 long splice variant to the response to ethanol was unknown.

To address the role of gamma 2 long in alcohol action, mice were genetically engineered such that at the molecular level they were unable to produce gamma 2 long without compromising production of gamma 2 short (Homanics et al., 1999). At the cellular level, electrophysiologic analysis of dorsal root ganglion neurons clearly demonstrated that in a neuronal context the gamma 2 long subunit is not absolutely required for ethanol to potentiate the effects of GABA at the GABA type A receptor. At the whole animal level, numerous ethanol-induced behaviors were compared between wild-type mice and gamma 2 long subunit knockout mice. Low-dose acute effects of ethanol such as anxiolysis, hyperlocomotion, and acute functional tolerance were not affected by the knockout. Acute, high-dose response to alcohol in a sleep-time assay similarly did not differ between genotypes. Dependence on alcohol, measured by quantifying the severity of the withdrawal reaction following termination of chronic exposure to ethanol, was also unaffected. Ethanol consumption and preference were not influenced by the lack of the gamma 2 long subunit (J. Wehner, unpublished observations). Together these results from the gamma 2 long subunit knockout mouse indicate that at the molecular, cellular, and whole animal levels, the gamma 2 long subunit is not the key to ethanol's actions at the GABA type A receptor and thus effort should be directed elsewhere.

3. ANESTHETICS

The primary effects of general anesthetics are very similar to alcohol. These drugs suppress excitatory neurotransmission and potentiate inhibitory neurotransmission. Whereas animal models have a long and productive history in anesthesia research, only recently have genetically engineered mice been applied to this field of endeavor. Several examples will be briefly presented below.

The beta 3 subunit of the GABA type A receptor is widely expressed in prenatal and neonatal brain. In adult brain, high levels of expression are maintained in cerebral cortex, hippocampus, and spinal cord. Thus, the beta 3 subunit is a major component of GABA type A receptor isoforms in these neuronal tissues. Because these specific tissues are likely involved in mediating the physiologic effects of anesthetics, these beta 3-containing receptors represent prime targets for the molecular mechanism of action of these drugs. To investigate the functional role of the beta 3 subunit of the GABA type A receptor, beta 3 subunit deficient mice were created by gene knockout technology (Homanics et al., 1997a). Many (~60%) beta 3 knockouts exhibited cleft palate, and most (~90%) of the beta 3-deficient mice died at birth for unknown reasons. However, the rare surviving mice were incredibly interesting. These mice exhibited a number of phenotypic abnormalities (e.g., epilepsy, cognitive impairment, poor motor skills, hyperactivity) that modeled the human genetic disorder Angelman Syndrome (DeLorey et al., 1998; Huntsman et al., 1999).

These mice have been extensively studied for behavioral responses to a wide variety of general anesthetics (Quinlan et al., 1998). Sensitivity to the volatile anesthetics halothane and enflurane was significantly reduced in knockout mice compared with controls in a test of the ability of the drugs to eliminate a purposeful response to a noxious stimulus, that is, the tail clamp/withdrawal assay. In contrast, response to these same two anesthetics in a loss-of-righting reflex assay was unchanged by the genetic alteration. Not only do these studies establish the importance of the beta 3 subunit in mediating the immobilization anesthetic end point, but these results also support the theory that different mechanisms mediate the different anesthetic end points (Eger et al., 1997). In addition, in a sleep-time assay in response to the injection of various anesthetic compounds, beta 3 knockout mice were found to be significantly less sensitive to etomidate and midazolam but were normally sensitive to pentobarbital and ethanol (Quinlan et al., 1998). These studies establish for the first time that at the whole animal level, anesthetic responses at the GABA type A receptor are subunit-selective. Furthermore, the selective alteration in response to some anesthetics but not to others casts serious doubt on the unitary theory of anesthesia that purports a single site of action for all anesthetic compounds. Instead, it is now apparent that individual anesthetic compounds can have very specific mechanisms of action that are not necessarily shared with all other anesthetic agents, although it appears that some compounds do indeed exhibit at least partial overlap.

Opiates are a clinically important adjunct to many anesthetic procedures. In fact, opiates can produce anesthesia by themselves. Of tremendous clinical significance is morphine, currently the most effective analgesic available. Opiates such as morphine exert their effects by interacting with the seven transmembrane domain opiate receptors, which are coupled to G-proteins. To date, three opiate receptors have been identified (mu, kappa, delta). The pharmacologic significance of the various subtypes of receptors has been the subject of much debate.

Gene knockout mice that lack a functional mu opioid-receptor gene have been produced and characterized (Loh et al., 1998; Matthes et al., 1996; Sora et al., 1997b; Tian et al., 1997). Pharmacologically, binding of the mu-selective ligand [^{3}H]DAMGO was completely abolished in knockout mice, confirming the specificity of the knockout of the mu receptor. Elimination of the mu receptor had minimal effects on the general health and well-being of the mice, although one study reported changes in host defense and reproductive systems (Tian et al., 1997). The opioid receptors that remained (delta, kappa) in the knockouts were present in normal amounts. Thus, these receptors were not upregulated to compensate for the absence of the mu receptor. In tests of thermal nociception in drug-naive animals, phenotypic differences were not detected in one study (Matthes et al., 1996) but in a separate study with a different line of mu-deficient mice, knockouts were significantly more sensitive to thermal stimuli (Sora et al., 1997b). At this time, it is not clear why the two studies differ on this point because they both clearly represent null alleles of the mu receptor. It is possible that differences in genetic background of the mice are confounding this issue (see below). These disparate results prevent a firm conclusion to be drawn concerning the involvement of endogenous opioids and the mu receptor for nociceptive perception. In contrast to this disagreement between studies, it was clear that morphine-induced analgesia was completely ablated in mu-receptor knockout mice (Matthes et al., 1996; Sora et al., 1997b). Additionally, knockout of the mu receptor eliminated the rewarding effects of morphine and prevented physical dependence on this drug (Matthes et al., 1996). It is surprising that the analgesic effects of the delta-specific agonists DPDPE and BUBU were nearly eliminated in the mu-deficient mice (Matthes et al., 1996; Sora et al., 1997a). This unexpected finding suggests that mu-receptor occupancy may be required for delta-specific agonists to exert their effects. In sum, knockout of the mu receptor conclusively establishes that this opioid receptor is the molecular target of morphine in vivo and eliminates the possibility that delta and kappa receptors contribute to the effects of this drug. Understanding morphine's mechanism of action at the molecular and whole animal levels should lead to improved methods of pain control that are devoid of deleterious side effects. Additional information on the mu-receptor knockouts, as well as knockouts of other opioid receptors and endogenous opioid peptides, can be found in a recent review (Kieffer, 1999).

4. CAVEATS

From the examples illustrated above, it is obviously clear that even though genetically engineered mice have only recently been applied to assist in the dissection of the molecular mechanism of action of alcohol and anesthetics, rapid and dramatic advances in our understanding are being made. However, conventional genetic engineering strategies, particularly global gene knockouts, are hampered by several limitations. Although these limitations do not discredit these strategies, one must exercise caution when interpreting results derived from their use. These limitations will be discussed briefly below. In the section that immediately follows, new genetic technologies that minimize or eliminate these shortcomings will be presented.

Animals derived from global gene knockout technology lack the gene of interest throughout embryonic development and all of adulthood in all cells of the body. The ubiquity of the knockout can create a number of potential problems. First, few genes act in isolation, especially in the nervous system. Absence of a gene that is normally expressed in the developing nervous system is thus likely to induce alterations in the level of expression of other genes. Thus, it is likely that this developmental compensation, which is secondary to the knockout, may confound interpretation of the phenotype. It may be very difficult to determine if the phenotype observed is directly due to the absence of the gene of interest or instead if the phenotype is due to compensatory changes.

For some studies, developmental compensation is not a major problem. For example knockout of the GABA type A receptor beta 3 subunit mentioned above was also created to model the human disease Angelman Syndrome (De-Lorey et al., 1998). Because in Angelman patients the beta 3 gene is globally absent, a conventional gene knockout is very appropriate. Nonetheless, whereas such a knockout mouse may indeed represent an accurate model of the human disease syndrome it is still difficult to sort out cause-and-effect because the nature and impact of the neurodevelopmental changes secondary to the knockout are entirely unknown. With respect to the impact of the beta 3 knockout on anesthetic responses (Quinlan et al., 1998), at present it is not clear if the altered anesthetic sensitivity is due to the specific absence of beta 3 or if the phenotype is due to compensatory changes.

A second illustration of this problem involves targeted inactivation of the alpha 6 subunit of the GABA type A receptor (Jones et al., 1997). Although unexpected, it was observed that the knockout of the alpha 6 subunit resulted in the concomitant selective loss of the delta subunit of the GABA type A receptor. Whereas the study revealed novel insight into subunit partnerships, it highlights the magnitude of unexpected changes in other gene products. Presently it is unclear how widespread this problem is because it is impossible to monitor expression of all genes that may be altered.

A second limitation of conventional knockout animals is that because the gene is knocked out in all cells of the animal, it is virtually impossible to ascribe the phenotypic changes to a particular neuronal circuit, brain region, or cell type. For example, the NPY knockout mentioned above displays an incredibly interesting change in alcohol consumption and sleep time (Thiele et al., 1998). Because NPY is expressed in cortex, amygdala, hippocampus, and other regions of the brain, one cannot determine which of these regions is critical for the behavioral alterations.

Lastly, a limitation that has recently received considerable attention from researchers concerns the genetic background of knockout animals (Gerlai, 1996; Banbury Conference on Genetically Engineered Mice, 1997; Smithies and Maeda, 1995). Due to technical limitations that are not understood, nearly all gene knockouts are produced by using Strain 129-derived embryonic stem cells. Because most substrains of Strain 129 are poor breeders, most knockouts are maintained on a hybrid genetic background by mating to another inbred strain, usually C57BL/6J. As illustrated in Figure 6.2, germline transmission of the targeted allele following mating of the Strain 129-derived chimeric animal with a C57BL/6J mouse results in F1-generation animals. These wild-type and heterozygous mice are genetically identical with the exception of the targeted allele; in all mice, one copy of each chromosome will be Strain 129-derived and one copy will be C57BL/6J-derived (Fig. 6.2B). However, upon crossing heterozygotes to produce mice of the F2 generation, meiotic recombination during gametogenesis shuffles regions of homologous chromosomes. The net result is F2 mice that differ genetically not only at the targeted mutation, but also throughout the genome (Fig. 6.2C). Note that each animal is unique; the origin of specific chromosomal segments is a random mix of sequences derived from either Strain 129 or C57BL/6J. Because parental inbred mouse strains often harbor polymorphic alleles that are capable of modifying the phenotype of interest, one must exercise caution when interpreting the phenotype observed. For polymorphic alleles that are not physically linked to the targeted mutation, this condition is only a minor problem, as these alleles will randomly segregate among mice of all genotypes. This heterogeneity in the genetic background can result in considerable animal-to-animal variation, which may mask small effects of the knockout. A far greater concern exists if the polymorphic genes are so close to the targeted locus that they are physically linked to the targeted mutation, that is, they are so close that recombination does not occur between them and the targeted gene. As illustrated in Figure 6.2C, the genes linked (denoted by hatched bar) to the targeted mutation will all be Strain 129-derived in the homozygous mutant animals. Whereas in the wild-type animals those genes linked to the wild-type allele will all be C57BL/6J-derived. Thus, it may be very difficult to conclusively assign the observed phenotype to the targeted mutation. It is conceivable that linked genetic polymorphisms are the underlying cause of the phenotype. Because one rarely knows what genes are linked to the gene of interest, the effect of linked genes is often overlooked. A recently published exception that clearly illustrates the problem is the demonstration that motor coordination on the rotorod in dopamine D2 receptor knockouts is strongly influenced by genetic background (Kelly et al., 1998). Solutions to this problem have been proposed (Gerlai, 1996; Mice, 1997; Smithies and Maeda, 1995), although there is no ideal solution that is fast, inexpensive, and straightforward. One strategy that is often used involves repeated back-crossing of mutant animals to an inbred strain. Although this technique can result in >99% genetic uniformity, the problem of linked genes remains (See Fig. 6.2D). It has been estimated that even after 12 generations of back-crossing, 300^+ linked genes may remain that are Strain 129-derived (Gerlai, 1996). Thus, back-crossing is not a satisfactory solution.

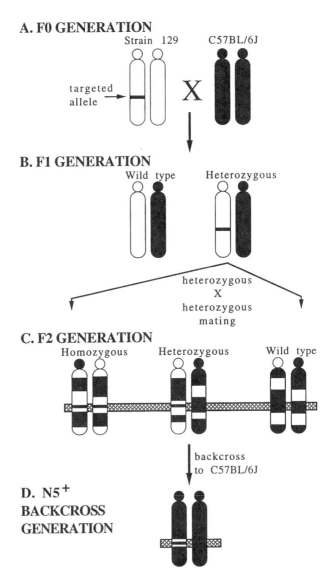

Figure 6.2. Parental origin of chromosomal segments can confound interpretation of gene targeting experiments. A. The founding (F0) generation consists of the Strain 129 derived germline chimera (chromosomes shown in white) and a C57BL/6J mouse (chromosomes shown in black). B. In the F1 generation, wild-type and heterozygous mice will be genetically identical with the exception of the targeted allele. C. Interbreeding of heterozygous mice results in an F2 generation consisting of homozygous, heterozygous, and wild type animals. During gametogenesis, crossover events that occur during meiosis shuffle Strain 129 and C57BL/6J derived chromosomal segments between homologous chromosomes. The net result is that each F2 animal possesses a unique mix of Strain 129 and C57BL/6J derived chromosomal segments. Note that the genes linked (indicated by hatched bar) to the targeted locus are all Strain 129 derived (white), whereas those genes linked to the wild type version of mutant allele are all C57BL/6J derived (black). D. Although backcrossing for greater than five generations ($N5^+$) can substantially reduce genetic heterogeneity, the problem of linked genes persists.

As described below, the conditional knockout approach will avoid the issue of linked genes confounding interpretation of a phenotype.

Again, it should be stressed that these limitations do not necessarily detract from the potential significance of the results; they only limit the extent of the conclusions that can be drawn from the experiments.

5. FUTURE DIRECTIONS

Genetically engineered mice have already had a dramatic impact on our understanding of how alcohol and anesthetics exert their mind-altering effects, despite the fact that this technology has only recently been applied to the problem. To date, nearly all relevant studies have relied upon either traditional transgenic animals or global gene knockouts. These approaches represent the first generation of genetic engineering technology. On the horizon are incredibly exciting novel second-generation approaches that will undoubtedly revolutionize the genetic engineering approach. The power of these techniques is that they provide the researcher with exquisite control over the extent and/or the timing of the genetic alteration. Consequently, experiments can be designed so that the limitations of conventional technologies are minimized or even eliminated. These novel techniques have already been demonstrated to be functional and are currently being unleashed on the problem of alcohol intoxication and anesthesia.

The recent evolution of the alcohol/anesthetic field is making it much more desirable to investigate the impact of subtle genetic alterations such as point mutations, in contrast to the global gene knockouts illustrated above. The groundwork for such studies has been laid by an elegant series of in vitro experiments from several different laboratories. Mihic et al. (1997) identified single amino acid residues of the GABA type A receptor that are absolutely critical for the effects of ethanol and volatile anesthetics at the GABA type A receptor. Similar studies have identified other amino acids that are critical for effects of other anesthetics on the GABA type A receptor (Belelli et al., 1997; Moody et al., 1997). However, because these studies were all conducted in vitro in model systems, the critical question of the importance of these amino acids to whole animal behavioral response to these drugs has not been addressed.

The so-called "gene knockin" approach will allow researchers to test the effects of these critical amino acids in vivo in intact animals. In contrast to a gene knockout in which the gene of interest is inactivated, a gene knockin results in a minor modification to the gene of interest. Often the change is limited to a single base pair mutation in the gene of interest while the remainder of the genome is unaltered.

Early gene knockins were created by using a laborious and technically demanding technique known as "hit and run" (Hasty et al., 1991) or "in-out" (Valancius and Smithies, 1991) gene targeting. This technique was used successfully by Limbird and colleagues (Lakhlani et al., 1997) to study anesthetic mechanisms. Specifically, they mutated the alpha-2a adrenergic receptor such that the aspartate residue at position 79 was changed to an asparagine. This precise, single genetic change eliminated whole animal behavioral responses to the alpha-2 agonist dexmetomidine. The knock-in animals were insensitive to dexmetomidine on tests of motor impairment, sleep time, anesthetic-sparing, and analgesia. Thus, this knock-in animal has helped to define the contribution of this subtype of the alpha-2 adrenergic receptor to the anesthetic effects of dexmetomidine.

Although the in-out technique has been successfully used on a number of occasions, it no longer is the method of choice for creating gene knock-ins. The recent application of site-specific recombinases to gene targeting has greatly facilitated the creation of subtle mutations. Details of this approach can be found elsewhere in this volume (Chapter 9). This methodology should have a huge impact on the alcohol/anesthetic mechanism fields in the near future, as it is being applied to determine the physiologic relevance of the alcohol and anesthetic sites recently identified in vitro (Belelli et al., 1997; Mihic et al., 1997; Moody et al., 1997).

A second approach to modifying genes, which should be a tremendous asset to the genetic engineers toolbox, is that of conditional gene knockouts. This technique also relies on the use of site-specific recombinase enzymes. The overall strategy for this methodology is illustrated elsewhere in this volume (Chapter 8). With this methodology, one can create gene knockout animals in which the gene of interest is deleted only in specific cells and/or at specific stages of the life of the animal. Because the gene is inactivated only in specific cells, it should be possible to assign phenotypic alterations in behavioral responses to alcohol and anesthetics to single genetic changes in defined neuroanatomic locations. Also by knocking out the gene of interest only after critical developmental events have passed, developmental compensation should be minimized or even eliminated. This approach will also be devoid of the confounding influence of linked genes. In these experiments, conditional knockout animals will be directly compared with animals that contain a targeted, but unrecombined and fully functional, gene of interest. These two sets of animals will be genetically identical, even for those chromosomal segments that are linked to the gene of interest. Under some circumstances, it may even be desirable to compare the same animal before and after recombination has occurred. Thus, the conditional knockout approach avoids the main limitations of conventional global knockout technology.

In the mammalian central nervous system, the conditional knockout approach was pioneered by Tonegawa and colleagues (Tsien et al., 1996a, 1996b). They specifically inactivated the NMDAR1 subunit of the glutamate receptor only in hippocampal pyramidal cells of weanling mice and demonstrated a selective effect on learning and memory. Similar experiments to conditionally inactivate many other genes are in progress in numerous labs around the world, and they should provide exciting insight into alcohol and anesthetic action.

Another rapidly emerging genetic technology is that of regulatable transgenes (see Chapter 2). Techniques are being developed that enable investigator control over the level and timing of transgene expression by simply adding or removing a drug from the animals' drinking water (Baron et al., 1999). Similar approaches use an injection of an inducer to control transgene expression. Such tight control over transgene expression will prevent issues of developmental compensation.

Perhaps the most exciting possibility on the horizon is that of combining several of the emerging technologies in one genetically manipulated animal model. For example, combining the regulatable transgene technology with the conditional gene knockout technology will enable the investigator to have direct control over the timing and the neuroanatomic location of the inactivation of the gene of interest. Such an experimental paradigm will avoid the limitations of currently used approaches, and they should allow for dramatic and exciting insight into the mechanism of action of alcohol and anesthetics.

REFERENCES

Banbury Conference on Genetically Engineered Mice (1997): Mutant mice and neuroscience: recommendations concerning genetic background. Neuron 19:755–759.

Baron U, Schnappinger D, Helbl V, Gossen M, Hillen W, Bujard H (1999): Generation of conditional mutants in higher eukaryotes by switching between the expression of two genes. PNAS USA 96: 1013–1018.

Belelli D, Lambert JJ, Peters JA, Wafford K, Whiting PJ (1997): The interaction of the general anesthetic etomidate with the gamma-aminobutyric acid type A receptor is influenced by a single amino acid. PNAS USA 94:11031–11036.

Cadoret, RJ. Genetics of alcoholism. In Alcohol and the Family: Research and Clinical Perspectives; Collins RL, Leonard KE, Searles JS, Eds. Guilford Press: New York, 1990; pp 39–78.

Carr LG, Foroud T, Bice P, Gobbett T, Ivashina J, Edenberg H, et al. (1998): A quantitative trait locus for alcohol consumption in selectively bred rat lines. Alcoh Clin Exp Res 22: 884–887.

Crabbe JC, Belknap JK, Buck KJ (1994): Genetic animal models of alcohol and drug abuse. Science 264:1715–1723.

Crabbe JC, Phillips TJ, Feller DJ, Hen R, Wenger CD, Lessov CN, et al. (1996): Elevated alcohol consumption in null mutant mice lacking 5-HT1B serotonin receptors. Nat Genet 14:98–101.

Crosby G, Marota JJ, Huang PL (1995): Intact nociception-induced neuroplasticity in transgenic mice deficient in neuronal nitric oxide synthase. Neuroscience 69:1013–1017.

DeLorey TM, Handforth A, Homanics GE, Minassian BA, Anagnostaras SG, Asatourian A, et al. (1998): Mice lacking the b3 subunit of the GABAA receptor have the epilepsy phenotype and many of the behavioral characteristics of Angelman Syndrome. J Neurosci 18:8505–8514.

Eger EI, Koblin DD, Harris RA, Kendig JJ, Pohorille A, Halsey MJ, et al. (1997): Hypothesis—Inhaled anesthetics produce immobility and amnesia by different mechanisms at different sites. Anesth Anal 84:915–918.

Engel SR, Lyons CR, Allan AM (1998): 5-HT3 receptor over-expression decreases ethanol self-administration in transgenic mice. Psychopharmacology 140:243–248.

Franks NP, Lieb WR (1994): Molecular and cellular mechanisms of general anaesthesia. Nature 367:607–614.

Gerlai R (1996): Gene-targeting studies of mammalian behavior: is it the mutation or the background genotype? Trends Neurosci 19:177–181.

Gunther U, Benson J, Benke D, Fritschy J, Reyes G, Knoflach F, et al. (1995): Benzodiazepine-insensitive mice generated by targeted disruption of the gamma 2 subunit gene of gamma-aminobutryic acid type A receptors. PNAS USA 92:7749–7753.

Harris RA, McQuilkin SJ, Paylor R, Abeliovich A, Tonegawa S,Wehner JM (1995): Mutant mice lacking the gamma isoform of protein kinase C show decreased behavioral actions of ethanol and altered function of gamma-aminobutyrate type A receptors. PNAS USA 92:3658–3662.

Hasty P, Ramirez-Solis R, Krumlauf R, Bradley A (1991): Introduction of a subtle mutation into the Hox-2.6 locus in embryonic stem cells. Nature 350:243–246.

Homanics GE, Delorey TM, Firestone LL, Quinlan JJ, Handforth A, Harrison NL, et al. (1997a): Mice devoid of γ-aminobutyrate type A receptor β3 subunit have epilepsy, cleft palate, and hypersensitive behavior. PNAS USA 94:4143–4148.

Homanics GE, Ferguson C, Quinlan JJ, Daggett J, Snyder K, Lagenaur C, et al. (1997b): Gene knockout of the alpha-6 subunit of the gamma-aminobutyric acid type A receptor: lack of effect on responses to ethanol, pentobarbital, and general anesthetics. Mol Pharmacol 51:588–596.

Homanics GE, Harrison NL, Quinlan JJ, Krasowski MD, Rick CEM, deBlas AL, et al. (1999): Normal electrophysiological and behavioral responses to ethanol in mice lacking the long splice variant of the γ2 subunit of the γ-aminobutyrate type A receptor. Neuropharmacology 38:253–265.

Homanics GE, Le NQ, Kist F, Mihalek RM, Hart AR, Quinlan JJ (1998a): Ethanol tolerance and withdrawal responses in GABA$_A$ receptor alpha 6 subunit null allele mice and in inbred C57BL/6J and Strain 129/SvJ mice. Alcoh Clin Exp Res 22:259–265.

Homanics GE, Quinlan JJ, Mihalek RM, Firestone LL (1998b): Alcohol and anesthetic mechanisms in genetically engineered mice. Front Biosci 3:d548–558.

Huntsman MM, Porcello DM, Homanics GE, DeLorey TM, Huguenard JR (1999): Reciprocal inhibitory connections control network synchrony in the mammalian thalamus. Science 283:541–543.

Ichinose F, Huang PL, Zapol WM (1995): Effects of targeted neuronal nitric oxide synthase gene disruption and nitroG-L-arginine methylester on the threshold for isoflurane anesthesia. Anesthesiology 83:101–108.

Jones A, Korpi ER, McKernan RM, Pelz R, Nusser Z, Makela R, et al. (1997): Ligand-gated ion channel subunit partnerships—GABA(a) receptor alpha(6) subunit gene inactivation inhibits delta subunit expression. J Neurosci 17:1350–1362.

Kelly M, Rubinstein M, Phillips T, Lessov C, Burkhart-Kasch S, Zhang G, et al. (1998): Locomotor activity in D2 dopamine receptor-deficient mice is determined by gene dosage, genetic background, and developmental adaptations. J Neurosci 18:3470–3479.

Kieffer BL (1999): Opioids: First lessons from knockout mice. Trends Pharmacol Sci 20:19–26.

Korpi ER, Koikkalainen P, Vekovischeva OY, Makela R, Kleinz R, Uusi-Oukari M, et al. (1999): Cerebellar granule-cell-specific GABAA receptors attenuate benzodiazepine-induced ataxia: evidence from alpha 6-subunit-deficient mice. Eur J Neurosci 11:233–240.

Lakhlani PP, MacMillan LB, Guo TZ, McCool BA, Lovinger DM, Maze M, et al. (1997): Substitution of a mutant alpha2a-adrenergic receptor via "hit and run" gene targeting reveals the role of this subtype in sedative, analgesic, and anesthetic-sparing responses in vivo. PNAS USA 94:9950–9955.

Loh HH, Liu HC, Cavalli A, Yang W, Chen YF, Wei LN (1998): mu Opioid receptor knockout in mice: effects on ligand-induced analgesia and morphine lethality. Mol Brain Res 54:321–326.

Maldonado R, Saiardi A, Valverde O, Samad TA, Roques BP, Borrelli E (1997): Absence of opiate rewarding effects in mice lacking dopamine D2 receptors. Nature 388:586–589.

Matthes HW, Maldonado R, Simonin F, Valverde O, Slowe S, Kitchen I, et al. (1996): Loss of morphine-induced analgesia, reward effect and withdrawal symptoms in mice lacking the mu-opioid-receptor gene. Nature 383:819–823.

McKernan RM, Whiting PJ (1996): Which GABAa-receptor subtypes really occur in the brain? Trends Neurosci 19:139–143.

Merikangas KR (1990): The genetic epidemiology of alcoholism. Psychol Med 20:11–22.

Mihic S, Ye Q, Wick M, Koltchine V, Finn S, Krasowski M, et al. (1997): Molecular sites of volatile anesthetic action on GABAA and glycine receptors. Nature 389:385–389.

Miyakawa T, Yagi T, Kitazawa H, Yasuda M, Kawai N, Tsuboi K, et al. (1997): Fyn-kinase as a determinant of ethanol sensitivity—relation to NMDA-receptor function. Science 278:698–701.

Moody EJ, Knauer C, Granja R, Strakhova M, Skolnick P (1997): Distinct loci mediate the direct and indirect actions of the anesthetic etomidate at GABA(A) receptors. J Neurochem 69:1310–1313.

Quinlan JJ, Homanics GE, Firestone LL (1998): Anesthesia sensitivity in mice lacking the β3 subunit of the GABA$_A$ receptor. Anesthesiology 88:775–780.

Rubinstein M, Phillips TJ, Bunzow JR, Falzone TL, Dziewczapolski G, Zhang G, et al. (1997): Mice lacking dopamine D4 receptors are supersensitive to ethanol, cocaine, and methamphetamine. Cell 90:991–1001.

Sallinen J, Link RE, Haapalinna A, Viitamaa T, Kulatunga M, Sjoholm B, et al. (1997): Genetic alteration of alpha 2C-adrenoceptor expression in mice: influence on locomotor, hypothermic, and neurochemical effects of dexmedetomidine, a subtype-nonselective alpha 2-adrenoceptor agonist. Mol Pharmacol 51:36–46.

Saudou F, Amara DA, Dierich A, LeMeur M, Ramboz S, Segu L, et al. (1994): Enhanced aggressive behavior in mice lacking 5-HT1B receptor. Science 265:1875–1878.

Sieghart W (1995): Structure and pharmacology of γ-aminobutyric acid$_a$ receptor subtypes. Pharmacol Rev 47:181–234.

Simonin F, Valverde O, Smadja C, Slowe S, Kitchen I, Dierich A, et al. (1998): Disruption of the kappa-opioid receptor gene in mice enhances sensitivity to chemical visceral pain, impairs pharmacological actions of the selective kappa-agonist U-50,488H, and attenuates morphine withdrawal. EMBO J 17:886–97.

Smithies O, Maeda N (1995): Gene targeting approaches to complex genetic diseases: atherosclerosis and essential hypertension. PNAS USA 92:5266–5272.

Sora I, Funada M, Uhl GR (1997a): The mu-opioid receptor is necessary for [D-Pen2,D-Pen5]enkephalin-induced analgesia. Eur J Pharmacol 324:R1–2.

Sora I, Takahashi N, Funada M, Ujike H, Revay RS, Donovan DM, et al. (1997b): Opiate receptor knockout mice define Mu receptor roles in endogenous nociceptive responses and morphine-induced analgesia. PNAS USA 94:1544–1549.

Thiele TE, Marsh DJ, Ste. Marie L, Bernstein IL, Palmiter RD (1998): Ethanol consumption and resistance are inversely related to neuropeptide Y levels. Nature 396:366–369.

Tian M, Broxmeyer HE, Fan Y, Lai Z, Zhang S, Aronica S, et al. (1997): Altered hematopoiesis, behavior, and sexual function in mu opioid receptor-deficient mice. J Exp Med 185:1517–1522.

Tsien JZ, Chen DF, Gerber D, Tom C, Mercer EH, Anderson DJ, et al. (1996a): Subregion- and cell type-restricted gene knockout in mouse brain. Cell 87:1317–1326.

Tsien JZ, Huerta PT, Tonegawa S (1996b): The essential role of hippocampal CA1 NMDA receptor-dependent synaptic plasticity in spatial memory. Cell 87:1327–1338.

Valancius V, Smithies O (1991): Testing an "In-Out" targeting procedure for making subtle genomic modifications in mouse embryonic stem cells. Mol Cell Bio 11:1402–1408.

Wafford KA, Burnett DM, Leidenheimer NJ, Burt DR, Wang JB, Kofuji P, et al. (1991): Ethanol sensitivity of the GABA-A receptor expressed in Xenopus Oocytes requires 8 amino acids contained in the γ2L subunit. Neuron 7:27–33.

Wafford KA, Whiting PJ (1992): Ethanol potentiation of GABA-A receptors requires phosphorylation of the alternatively spliced variant of the γ2 subunit. FEBS Lett 313:113–117.

CHAPTER 7

IN VIVO FUNCTIONS OF THE c-*MET* GENE: (HGF/SF RECEPTOR)

SILVIA GIORDANO and PAOLO M. COMOGLIO

1. INTRODUCTION

The *Met* gene encodes the tyrosine kinase receptor for the cytokine hepatocyte growth factor/scatter factor (HGF/SF), which displays distinctive biological properties. In addition to promoting cell growth and protection from apoptosis, activation of the HGF/SF receptor results in the control of cell dissociation, migration into extracellular matrices, and a unique process of differentiation called "branching morphogenesis." Through the concerted regulation of these complex phenomena, the *Met*-HGF/SF couple plays an important role throughout embryogenesis, regeneration, and reconstruction of normal organ architecture.

2. HEPATOCYTE GROWTH FACTOR/SCATTER FACTOR

HGF/SF owes its double denomination to the independent discovery of its abilities to induce growth of hepatocytes and dissociation/motility of epithelial cells (scattering) (Gherardi et al., 1989; Gohda et al., 1988; Nakamura et al., 1986; Stoker et al., 1987; Weidner et al., 1990; Zarnegar and Michalopoulos, 1989). After biochemical purification and cDNA cloning, it was demonstrated that HGF and SF are the same molecule (Miyazawa et al., 1989; Nakamura et al., 1989; Naldini et al., 1991; Weidner et al., 1991).

HGF/SF expression starts during embryogenesis and by day E18 it is widely distributed in developing epithelia, limb buds, and neural tissues (Sonnenberg et al., 1993; Thery et al., 1995; Andermarcher et al., 1996). In the adult stage, several mesenchymal-derived cells secrete HGF/SF, while the specific receptor is expressed on target cells, mainly of epithelial type. Although HGF/SF is normally found in serum, where its concentration increases in certain pathological conditions, it is thought to work predominantly in a paracrine mode

Genetic Manipulation of Receptor Expression and Function,
Edited by Domenico Accili.
ISBN 0-471-35057-5   Copyright © 2000 Wiley-Liss, Inc.

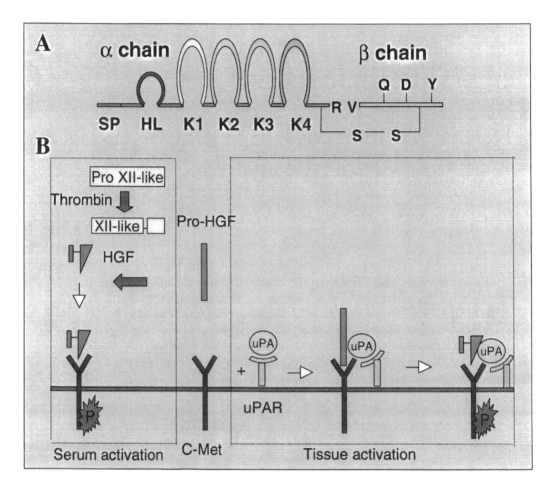

Figure 7.1. (A) A diagram of the structure of HGF/SF, a heterodimer consisting of a heavy α chain and a light β chain held together by a disulfide bond. The mature heterodimer is formed by proteolytic digestion at a specific dibasic arginine-valine (R-V) site. Starting from the amino terminal signal peptide (SP), the α chain contains a hairpin loop (HL) followed by four kringles (K), whereas the β chain contains a serine-protease-like structure. The lack of proteolytic activity in the HGF/SF molecule is due to the replacement of the histidine (H) and serine (S), residues lying within the catalytic site of conventional serine proteases, with glutamine (Q) and tyrosine (Y), respectively. (B) Schematic model of the activation pathways for the HGF/SF single-chain precursor. HGF/SF can undergo activation by uPA at tissue sites (right panel). UPA forms a stable complex with pro-HGF/SF both in the extracellular matrix and on the cell membrane. As uPA remains bound to HGF/SF, the yield of two-chain factor depends on the amount of active uPA available. Pro-HGF/SF can also be processed by serum-derived convertases, acting as soluble catalysts in the extracellular space after tissue injury (left panel). (See color plates.)

(Sonnenberg et al., 1993). In fact, HGF/SF is a heparin-binding factor, naturally accumulating in the extracellular matrix and in basement membranes,
bound to heparan sulphate proteoglycans (HSPGs) (Kobayashi et al., 1994).

HGF/SF is a large molecule ($\sim$100 kDa), similar to the serine-proteases of
the blood-clotting cascade (40% homology to plasminogen) but devoid of enzymatic activity. It is a heterodimer consisting of a 62-kDa α and a 32/34-kDa
β chain, held together by a disulfide bond (Fig. 7.1A). The α (heavy) subunit
contains four typical "kringle domains", (an 80-amino-acid, double-looped
structure formed by 3 internal disulfide bridges), similar to those observed in
the coagulation factors. Although their exact function is presently unclear,
kringles are thought to act as protein-protein interaction motifs. The α chain
also contains the binding domains for the receptor and for HSPGs (Mizuno et
al., 1994). It has been shown that the low-affinity binding site for HSPGs, although important for ensuring sequestration of the factor in the proximity of
target cells, is not essential for HGF/SF binding to *Met* receptor and for signaling. The β (light) subunit of HGF/SF is closely related to the catalytic domain
of serine proteases; however, the serine residue of the active site is substituted
with a tyrosine. Therefore, HGF/SF shares structural homology and activation
mechanism with plasminogen but has lost protease activity during evolution
(for an overview, see Donate et al., 1994).

HGF/SF is secreted as a single-chain, biologically inert glycoprotein precursor (pro-HGF/SF). Under appropriate conditions, pro-HGF/SF is converted into
its bioactive form by a proteolytic cleavage between the two positively charged
amino acids Arg^{494}-Val^{495} (Fig. 7.1B). Four proteases are reported to activate
HGF/SF in vitro: Urokinase-type (uPA) and tissue-type (tPA) plasminogen activators (Naldini et al., 1992; Mars et al., 1993); a serine protease isolated from
serum, homologous to coagulation factor XII (Miyazawa et al., 1993); and coagulation factor XII itself (Shimomura et al., 1995). It has also been shown that
certain epithelial cells secrete two potent inhibitors of proHGF/SF activation
(HAI-I and HAI-II), which might control this process (Kawaguchi et al., 1997).
Thus, the processing of proHGF/SF involves a complex pathway, including activators and inhibitors. Nevertheless, it is unlikely that activation of proHGF/SF
is the rate-limiting step for the availability of this factor because transgenic mice
overexpressing proHGF/SF under the control of a strong metallothionein promoter display a phenotype consistent with an excess of HGF/SF activity
(Takayama et al., 1996). This finding implies that the activation system has a
high capacity and can efficiently convert surplus amounts of this growth factor.

3. THE HGF/SF RECEPTOR (C-*MET*)

HGF/SF binds to a specific cell membrane receptor belonging to the tyrosine
kinase receptor superfamily. The HGF/SF receptor, encoded by the c-*Met*
proto-oncogene (Cooper et al., 1984; Giordano et al., 1989a; Bottaro et al.,
1991; Naldini et al., 1991), is a single-pass cell membrane tyrosine kinase. It
is a dimeric molecule composed of a 50-kDa (α) chain disulfide linked to a
145-kDa (β) chain in an α / β complex of 190 kDa. The α chain is exposed at
the cell surface, and the β chain spans the plasma membrane (Fig. 7.2).

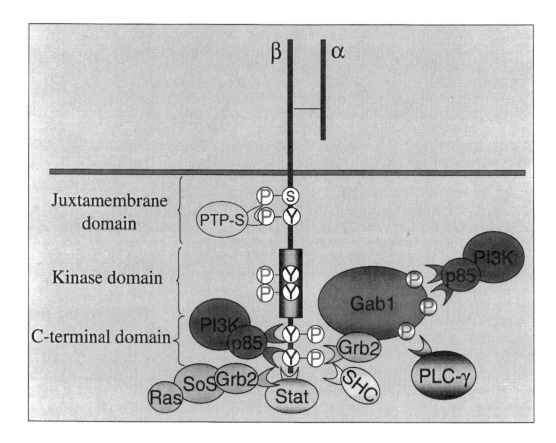

Figure 7.2. Schematic representation of the structure of the *Met* receptor. The receptor is a single-pass, disulphide-linked α/β heterodimer . The α chain is an extracellular gly-coprotein, the β chain is a transmembrane subunit responsible for the tyrosine kinase ac-tivity. The intracellular domain of the receptor includes a tyrosine kinase catalytic site (grey box) flanked by distinctive juxtamembrane and carboxy-terminal sequences. Two phosphorylated tyrosine residues contained within the kinase domain have a positive regulatory effect on the enzyme activity, whereas a serine and a tyrosine residue in the juxtamembrane domain negatively regulates the kinase. The carboxy-terminal portion includes two tyrosine residues that, when phosphorylated, form a specific docking site for multiple signal transducers and adaptors. (See color plates.)

The HGF/SF receptor is synthesized as a large precursor (pr170) that in-cludes both the α and β chains, undergoes co-translational glycosylation, and is cleaved by proteases of the "furin" family to form the mature subunits (Gior-dano et al., 1989; Mark et al., 1992; Crepaldi et al., 1994). Both α and β sub-units are necessary for the biological activity.

The intracellular portion of the receptor can be divided into three functional domains: (1) a juxtamembrane domain; (2) a tyrosine kinase catalytic domain; and (3) a C-terminal tail. The juxtamembrane domain negatively regulates *Met*-mediated activity. Two residues are responsible for the inhibitory effects of the juxtamembrane region: Ser^{975} and Tyr^{1003}. The negative effect of Ser^{975} depends on its phosphorylation, which results from activation of protein kinase-C and of a Ca^{2+}-calmodulin-dependent kinase (Gandino et al., 1990; Gandino et al., 1994). The negative effect exerted by Tyr^{1003} is less understood and may involve phosphorylation-dependent recruitment to the receptor of a cytosolic tyrosine phosphatase (Villa Moruzzi et al., 1993).

The tyrosine kinase catalytic domain contains the major phosphorylation site, represented by the tyrosine residues 1234 and 1235, which are essential for full activation of the enzyme (Ferracini et al., 1991; Longati et al., 1994). Upon phosphorylation of these residues the enzymatic activity of the *Met* kinase is strongly up-regulated in an autocatalytic fashion (Naldini et al., 1991).

The C-terminal tail domain of the *Met* receptor is crucial for its biological activity: it consists of a short sequence containing two tyrosines that become phosphorylated upon HGF/SF binding and is alone responsible for mediating high-affinity interactions with multiple SH2-containing cytoplasmic effectors. Unlike the docking sites identified in other tyrosine kinase receptors, the *Met* sequence $Y^{1349}VHV$NAT$Y^{1356}VNV$ can interact with GRB-2, p85-PI3K, PLCγ, STAT3, SHC, and GAB1 and is therefore referred to as the HGF/SF receptor "multifunctional docking site" (Ponzetto et al., 1994; Pelicci et al., 1995; Weidner et al., 1996; Boccaccio et al., 1998). GRB-2, which has a strong requirement for asparagine in the $+2$ position, specifically interacts only with the sequence Y^{1356}VNV (Ponzetto et al., 1996).

The "multifunctional docking site" is essential for *Met*-mediated biological responses. Substitution of both tyrosines with phenylalanine does not affect the receptor kinase activity but completely abolishes proliferation, motility, invasion, and tubulogenesis (Ponzetto et al., 1994). Recently, genetic evidence confirmed the crucial role of the "multifunctional docking site" for *Met* signaling in vivo. A mutated receptor carrying the tyrosine to phenylalanine mutations was introduced by homologous recombination into the *Met* locus of the mouse genome (Maina et al., 1996). Homozygous mice died in utero at a similar time compared with the *Met* null mice and had the same defects (see after), meaning that the "multifunctional docking site" is essential to transduce *Met* signals during mouse development.

The components of the signal-transduction machinery implicated in individual *Met*-mediated biological responses have been partially elucidated. Coupling of the receptor to the *Ras*–MAP kinase pathway is both essential and sufficient for proliferation (Ponzetto et al., 1996; Giordano et al., 1997). Conversely, activation of PI3 kinase as well as the Ras-Rac/Rho pathways is required to induce motility (Ridley et al., 1995; Royal and Park, 1995). Notably, the concomitant activation of *Ras*–MAP kinase and PI-3K is both necessary and sufficient for the HGF/SF receptor invasion (Giordano et al., 1997; Bardelli et al., 1999). The morphogenetic response, which is specifically induced by *Met* and cannot be elicited by other receptor tyrosine kinases (Sachs M. et al., 1996), involves Gab1 and STAT3 (Weidner et al., 1996; Boccaccio et al., 1998). The relationship between these two molecules is not clear, even if experimental evidence suggests that Gab1 is directly involved in the activation of STAT3. These data show that *Met* can activate multiple signal transduction pathways and that the elicited biological responses depend on activation of specific transducers. The availability of the molecules in different cell types can account for the different biological responses observed.

4. *MET*-RECEPTOR ACTIVATION TRIGGERS "INVASIVE GROWTH"

As mentioned, HGF/SF was initially identified as a growth factor for hepatocytes and the "scattering" response was considered specific for a subset of epithelial cells (e.g., MDCK). Later it became clear that both the mitogenic action and the scatter effect are different facets of a complex biological response—called "invasive growth"—that follows activation of the same receptor. Invasive growth results from the activation of a specific genetic program leading to invasion and metastasis in pathological conditions but essential to allow cell migration into extracellular matrices in morphogenetic processes during development. The following paragraphs illustrate the processes induced by *Met* receptor activation on cultured cells in vitro and during embryo development in vivo.

5. CONTROL OF CELL GROWTH AND PROTECTION FROM APOPTOSIS

Growth of primary hepatocytes, which normally cannot be maintained in cell culture for long periods, is clearly stimulated by HGF/SF. It is interesting that transgenic expression of HGF/SF in liver parenchymal cells leads to increased organ growth (Shiota et al., 1994). A growth response to HGF/SF is also observed in several other epithelial cells, in melanocytes and in endothelial cells, all expressing the *Met* receptor (Kan et al., 1991; Matsumoto et al., 1991; Bussolino et al., 1992). Moreover, HGF/SF acts as a survival and regeneration factor after severe damage to epithelial organs such as renal tubular failure, acute hepatic failure or partial hepatectomy, and acute lung injury (Nagaike et al., 1991; Yanagita et al., 1993).

HGF/SF protects epithelial cells from apoptosis induced in several experimental conditions. This ability has been clearly demonstrated in the case of hepatocytes. Targeting of an activated *Met* to the liver renders hepatocytes resistant to apoptotic stimuli (such as anoikis, drugs, and anti-FAS antibodies) and is permissive for immortalization of hepatocytes (Amicone et al., 1996).

A protective role of HGF/SF for hepatocytes has been shown also by knockout experiments. Mice lacking HGF/SF fail to complete development and die in utero (Schmidt et. al., 1995). One of the more affected organs is the embryonic liver, which is reduced in size and shows extensive loss of parenchymal cells. Histological analysis shows enlargement of sinusoidal spaces and dissociation of parenchymal cells. The dissociated cells often show signs of apoptosis, including nuclear fragmentation. A key physiological role of HGF/SF in liver development is also supported by results on ES cells carrying two targeted alleles of the *Met* gene: c-*Met* -/- ES cells cannot contribute to the liver but contribute to various other organs (Bladt et al., 1995).

6. SCATTERING AND EPITHELIAL MORPHOGENESIS

The scatter effect of HGF/SF can be observed "in vitro" in several epithelial cells having a flat polygonal phenotype and growing in islets. When treated

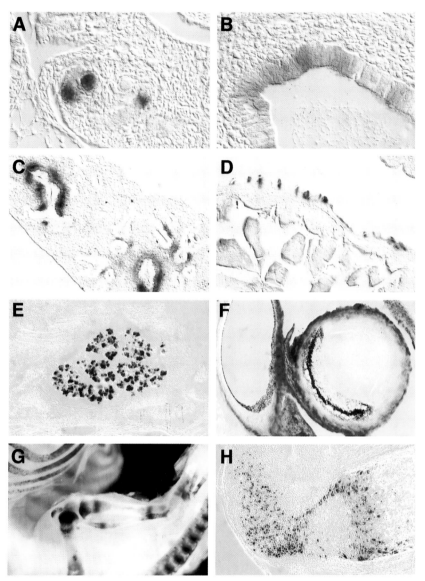

Figure 5.2. See page 73 for full caption.

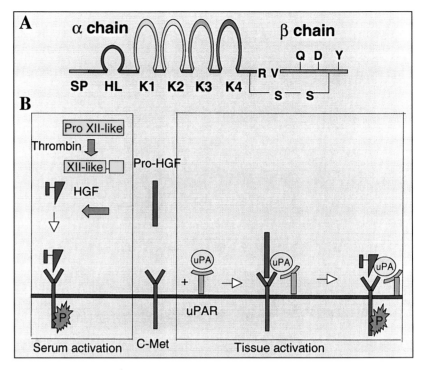

Figure 7.1. See page 112 for full caption.

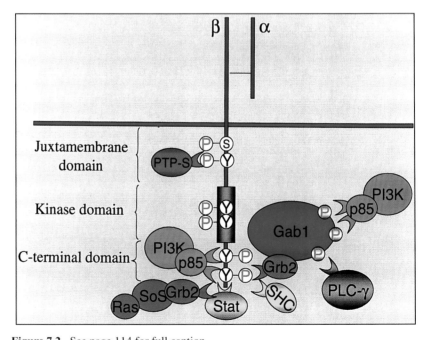

Figure 7.2. See page 114 for full caption.

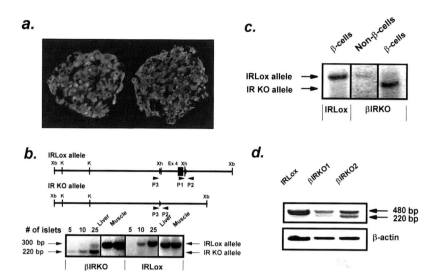

Figure 8.8. See page 144 for full caption.

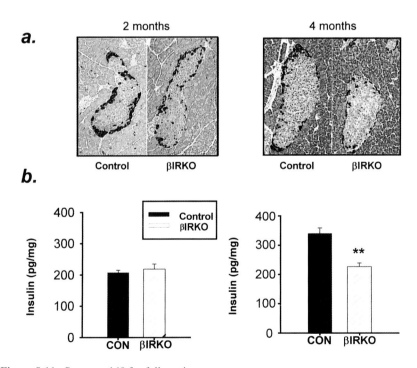

Figure 8.11. See page 149 for full caption.

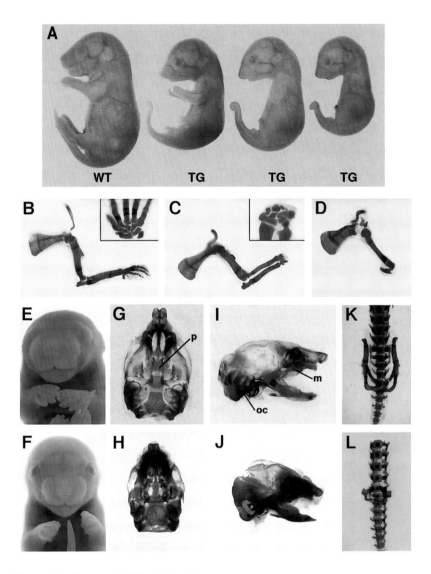

Figure 9.3. See page 165 for full caption.

with HGF/SF, the cells dissociate, form pseudopodia, and move on the plastic surface ("scattering") (Stoker et al., 1987). "In vivo", epithelial cells are usually packed in sheets with firm junctional complexes and are often polarized. Notably, the *Met* receptor is expressed on the basolateral surface (Crepaldi et al., 1994b), in contact with the basement membrane where HGF/SF secreted by stromal cells accumulates. Complex morphogenetic processes have been reconstituted in vitro by growing epithelial cells in collagen tridimensional gels. Under these conditions, most cells form spheroids and become susceptible to growth arrest and apoptosis, unless protected by the presence of survival signals. When HGF/SF is added under these conditions, it protects from apoptosis and stimulates proliferation; cell sprouts bud from the spheroids and eventually polarize to form branching tubules (Montesano et al., 1991; Medico et al., 1996). In the developing kidney, HGF/SF and *Met* are highly expressed at day 11.5, corresponding to the onset of tubulogenesis and branching morphogenesis (Santos et al., 1994). Defects in the HGF/SF-*Met* pathway may contribute to cystic diseases of epithelia, such as polycystic kidney disease.

Epithelial cells derived from different tissues (colon, mammary gland, kidney, prostate, and lung) show distinct differentiative paradigms when grown in collagen gels in the presence of HGF/SF (Brinkmann et al., 1995). This factor has also been involved in organogenesis of liver and pancreas, priming the differentiation of their respective progenitor cells (Hu et al., 1993; Calvo et al., 1996; Jeffers et al., 1996). HGF/SF thus appears to act as an "inductive", rather than "instructive," factor triggering cell type-specific differentiation programs, which have in common transient migration and eventual polarization of cells that form differentiated tridimensional structures. Embryonic organs cultured "in vitro" can undergo morphogenic and differentiation steps mimicking their development "in vivo". Inhibition of HGF/SF activity often leads to developmental abnormality (Yang et al., 1995; Tabata et al., 1996). The embryonic death of HGF/SF and c-*Met*-null embryos has precluded analysis of developmental events, such as morphogenesis of the mammary gland epithelium, which occurs later in development. Other processes, for example kidney or lung branching, appear to be unaffected in these mice, suggesting that compensatory mechanisms are at work in the embryo.

7. CONTROL OF CELL MIGRATION IN EARLY EMBRYOGENESIS

During gastrulation, HGF/SF and its receptor are both selectively expressed in the endoderm and in the mesoderm (Andermarcher et al., 1996). In early organogenesis, co-expression of HGF/SF and *Met* is found, for example, in neural crest cells and is probably responsible for the migration of cells into peripheral tissues. Subsequently, the expression of HGF/SF becomes limited to the mesenchymal cells, whereas that of the receptor is induced in the surrounding ectoderm. At this stage (approx. E13), HGF/SF contributes to the development of several epithelial organs (e.g., lungs, liver, and gut) and induces migration of specific myogenic precursor cells (Andermarcher et al., 1996). This observation was strengthened by analysis of c-*Met* gene targeting in mice. In c-*Met* homozygous mutant (-/-) mouse embryos, Bladt and colleagues observed that the limb bud and diaphragm are not colonized by myogenic precursor cells and, as a consequence, skeletal muscles of the limb and diaphragm do not develop (Bladt et al., 1995). In physiological conditions, HGF/SF is expressed in limb mesenchyme and can thus provide the signal for migration, which is received by c-*Met* expressing myogenic precursors. The effect of the mutation and the expression pattern are consistent with a role of c-*Met* and HGF/SF as key regulator molecules in the migration of myogenic cells. HGF/SF of the limb thus provides a spatially defined signal that induces the migration of myogenic precursor cells and may also direct migrating cells to the target. It is also interesting to note that to allow the migration of the myogenic precursors, the epithelium lining the lateral plate of dermomyotomes must be dissociated by the action of HGF/SF (Brand-Saberi et al., 1996). Furthermore, "Splotch" mice harboring mutations in a gene regulating c-*Met* expression (named Pax-3) show similar major disruptions in limb muscle development. This finding is probably due to the loss of c-*Met* expression in the myogenic precursor cells of the somitic dermomyotome, resulting in failure of their migration in the limb bud (Yang et al., 1996).

The critical role of tightly regulated HGF/SF expression for correct development is shown also by the analysis of transgenic mice overexpressing HGF/SF. These animals exhibit ectopic skeletal muscles and melanocytes in the central nervous system. This condition suggests that HGF/SF has scatter activity "in vivo" and can function as a morphogenetic factor, regulating migration and/or differentiation of selected populations of myoblasts and neural crest cells (Takayama et al., 1996).

8. MESENCHYMAL-EPITHELIAL INTERACTIONS AND EMBRYONAL TUBULOGENESIS

Epithelia and mesenchyme interact during various physiological processes, and exchange of signals between these compartments has been recognized as a major driving force in epithelial morphogenesis. Several observations indicate an important role for c-*Met* and HGF/SF in this process, and one of the best examples is the formation of kidney tubules from metanephric tissue and ureteric buds. The metanephric mesenchymal cells usually secrete HGF/SF but do not express its receptor. In contrast, the epithelial cells of the ureteric bud express the receptor only. When the latter make contact with the metanephros, they form branching tubules; in turn, some of the metanephric cells start to express the receptor for HGF/SF and—stimulated by the autocrine loop—undergo a process known as "mesodermic-epithelial transition." They acquire epithelial features, form cell-cell junctions, polarize, and eventually form tubules (Sonnenberg et al., 1993). It is interesting to note that, although all the mesenchymal cells of the renal cortex initially express HGF/SF, the receptor is found only in those that give origin to the renal tubules and not in the cells forming the glomeruli. In the adult, co-expression of HGF/SF and its receptor persists uniquely in mesangial cells, which probably represent a poorly differentiated compartment for regeneration (Kolatsijoannou et al., 1995).

The mammary gland is a good example of an organ that undergoes major morphological changes after birth and where branching development takes place during pregnancy and lactation (Pepper et al., 1995). This complex differentiation program relies on sequential stimulation of the epithelium by HGF/SF and neuregulin, secreted by the adjacent mesenchyme in distinct developmental stages (Yang et al., 1995). HGF/SF promotes the branching of the ductal tree and is expressed at puberty, whereas neuregulin is responsible for the alveolar budding and the production of milk proteins during pregnancy and lactation.

9. INTRODUCTION OF POINT MUTATIONS IN THE GENOME BY HOMOLOGOUS RECOMBINATION: THE KNOCK-IN TECHNIQUE

Because inactivation of either HGF or *Met* gene in the mouse causes embryonal lethality between E12.5 and E15.5 due to abnormal development of the placenta, it is not possible to derive from these studies information about the role played by these two genes during development of different tissues and organs. New techniques are now available to circumvent this problem. One of these is based on the principle of gene inactivation in selected cell types by making use of site specific recombinases—for example *cre* from bacteriophage P1. To create mice that lack the specific gene in certain tissues only, the gene must be knocked out not in embryonal stem (ES) cells but later on in cells of the developing embryo or even of the adult mouse. This process results in survival of the mice and allows studies of the role of the specific gene in the target tissues (Gu et al., 1994; Kuhn et al., 1995). Alternatively, to insert point mutations in the target gene, the so-called "knock-in" technique has been developed. Various procedures have been developed to repeatedly alter a target gene (Lewis et al., 1996; Hasty et al., 1991; Liu et al., 1998) The general principle is based on a two-step gene-targeting strategy. The first targeting step introduces a selectable marker into the genome of the ES cells by conventional gene targeting. The second step introduces additional modifications while at the same time deleting the marker, which was inserted at the first step. This second event is thus detected by selecting against the initial marker gene. Such a technique was used, for example, by Detloff and colleagues to target the mouse β globin gene, to substitute it with human mutated globin genes (Detloff et al., 1994).

10. NEW INSIGHT IN *MET* FUNCTIONS FROM KNOCK-IN MICE

As previously described, although considerable progress has been made in understanding the mechanisms of *Met* signaling in cultured cells, the lethality of either *Met* or HGF/SF knock-out mice prevented further studies on the role of these genes in development and function of different tissues and organs. For this purpose, taking advantage of what is known about *Met* signal transduction, Maina and colleagues decided to construct "knock- in" mice using a *Met* receptor impaired in its signal-transduction properties. It is known that in cellular systems the biological activity of the *Met* receptor depends on the presence of two phosphotyrosines ($Y^{1349}VHV$NAT$Y^{1356}VNV$) in the carboxy-terminal tail, which act as a "multifunctional docking site" for multiple SH2-containing signal transducers (Ponzetto et al., 1994). The $Y^{1356}VNV$ has the unique ability, compared with

$Y^{1349}VHV$, to bind also Grb2, which directly links the receptor with Ras. Mutation of Y^{1356} alone interferes heavily with all *Met*-mediated events, whereas mutation of Y^{1349} has only a limited effect on transformation and no effect on motility. Mutation of both tyrosines completely abolishes *Met* function (Ponzetto et al., 1994; Weidner et al., 1995; Ponzetto et al., 1996). The contribution of the Ras pathway in the HGF/SF responses has been studied with a *Met* mutant in which Grb2 binding is specifically abrogated while all the other effectors can still bind. This condition has been achieved by substituting the asparagine in position $+2$ of Y^{1356} with a histidine. Analysis of this mutant in cell cultures has shown that a direct link with Grb2 is required to promote transformation but is not essential to trigger the scatter response (Ponzetto et al., 1996; Giordano et al., 1997).

To discern if what has been observed in vitro with these mutants can be found also in vivo, Maina and colleagues aimed at interfering with *Met* receptor coupling to its effectors in mice (Maina et al., 1996). Among the several "knock-in" approaches available, the authors followed the one utilized by Hanks and colleagues, who made use of it to evaluate the functional equivalence of En-1 and En-2 genes (1995). These authors introduced an En-2 gene into the En-1 locus by homologous recombination. The result showed that the endogenous En-1 gene was inactivated (En-1 knock-out) but, at the same time, the En-2 gene introduced within it was controlled by En-1 regulatory sequences and expressed as an En-1 product. With the same approach, Maina constructed a targeting vector containing a human *Met* cDNA fragment coding for the transmembrane and the cytoplasmic domain of the receptor and inserted it in-frame, as a cassette, in a mouse genomic clone. Following the recombination event, the resulting *Met* protein encoded by targeted locus consists of a murine extracellular domain fused to a human transmembrane and cytoplasmic domain. With this technique, three different constructs have been produced: (1) a wild-type human *Met* cDNA carrying no mutation in the intracellular domain to have a control mouse; (2) a so-called $Met^{D/D}$ cDNA containing a double Y-to-F substitution (Y^{1349}VHVNATY^{1356}VNV to F^{1349}VHVNATF^{1356}VNV) to create an inactive signaling mutant; (3) the $Met^{Grb2/Grb2}$ cDNA, carrying N-to-H substitution (Y^{1349}VHVNATY^{1356}V*N*V to Y^{1349}VHV-NATY^{1356}V*H*V), selectively abrogating the Grb2 binding site, to produce a partial loss-of–function *Met* mutant. Screening of selected ES cells was performed by polymerase chain reaction (PCR) and Southern blotting. As shown in Figure 7.3, panel B, to perform PCR the authors chose two oligos: one recognizing sequences inside the targeting vector and one corresponding to genomic sequences located downstream from the 3' end of the vector. As expected, an amplified product could be observed only in ES cells that had undergone homologous recombination. Further analysis was performed by Southern blotting on genomic DNA obtained by the same ES cells. Genomic DNA was digested with NdeI and probed with the sequence indicated in the figure. As shown, the probe used identifies a 16-Kb fragment in the wild-type allele and a 12-Kb fragment in the recombinant allele (an additional NdeI site is present in the inserted cDNA). Targeting of both alleles thus results in the presence of a single 12-Kb band, whereas the appearance of both a 12- and a 16-Kb band reveals heterozygosity.

Loss of the *Met* multifunctional docking site ($Met^{D/D}$ construct) resulted in embryonic lethality and produced a phenotype essentially coinciding with that observed in *Met* and HGF null mutants. In these embryos the size of the liver

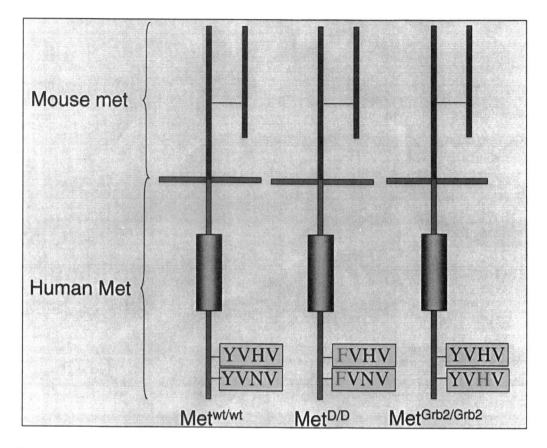

Figure 7.3. Panel A Schematic representation of the structure of the *Met* receptor in "knock-in" mice. A human *Met* cDNA fragment coding for the transmembrane and the cytoplasmic domain of the receptor was inserted in-frame into a mouse genomic clone. The resulting protein consists of a murine extracellular domain (blu) fused to a human transmembrane and cytoplasmic domain (red). The intracellular domain includes a tyrosine kinase catalytic site (grey box) and the multifunctional docking site (YVHV/YVNV). The three constructs are (1) left: a control construct carrying no mutation in the intracellular domain; (2) middle: a cDNA containing a double Y → F substitution to create an inactive signaling mutant (*Met*[D/D]); (3) right: a construct carrying N → H substitution, selectively abrogating the Grb2 binding site (*met*[Grb2/Grb2]).

(continued)

was reduced and the limbs, tip of the tongue, and diaphragm lacked muscles entirely. In contrast to embryonal lethality of *Met*[D/D] embryos, alive *Met*[Grb2/Grb2] mice were obtained. In these animals muscles deriving from migratory precursor where heavily affected, indicating a requirement for Grb2-mediated signaling in migratory myoblasts. This mutation does not affect *Met* ability to influence cell motility because mutant myoblasts leave the somites and begin entering the limb bud; this behavior is in agreement with the data collected "in vitro," showing that cells expressing the same mutant display a normal scatter response, (Ponzetto et al., 1996). What seems to be defective in *Met*[Grb2/Grb2] mouse

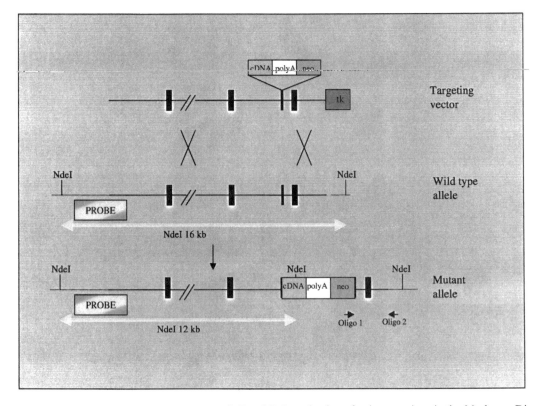

Figure 7.3. (*continued*) Panel B. Introduction of point mutations in the *Met* locus. Diagram of the targeting vector, wild-type, and mutant alleles. The human *Met* cDNA fragment is fused in-frame with the third exon of the genomic clone. After homologous recombination, the neo cassette remains within the mutant allele. The 5' probe identifies a 16-kb NdeI DNA fragment in the wild-type allele and a 12-kb NdeI DNA fragment in the mutant allele. Oligonucleotides used for PCR screening of the selected ES cells are shown. (Modified form Maina et al., 1996)

myoblasts is the ability to survive and proliferate while reaching their final destination. This, again, is in keeping with the "in vitro" data showing that Ras activation is essential for cell proliferation.

Another interesting observation derived from these mice is that *Met* is required for the development of secondary muscle fibers. *Met*[Grb2/Grb2] mice, in fact, show a 28% reduction of secondary fibers, probably due to a defect in proliferative activity of fetal myoblasts later in embryogenesis.

Other than elucidating the role of *Met* in controlling migration and growth of muscle precursor cells, these engineered mice have been very useful in trying to understand the role played by *Met* in the development of the nervous system. A possible role for this receptor was first suggested by the fact that both HGF/SF and *Met* are expressed in the developing nervous system and by the observation that HGF/SF promotes neural induction (Stern et al., 1990), it stimulates Schwann cell proliferation (Krasnoselsky et al., 1994), and it promotes axon outgrowth from P19 embryonal carcinoma cells (Yang and Park, 1993; Yamagata et al., 1995). Moreover, analysis of HGF/SF or c-*Met* null mice showed defects in limb motor axon branching. Ebens and colleagues clearly identified HGF/SF as a limb mesenchyme-derived chemoattractant for motor axons and, at later stages, as a muscle-derived survival factor for motoneurons (1996).

A further role for *Met* receptor signaling in the development of the nervous system was discovered by analyzing mice expressing the above-cited HGF receptor signaling mutants. Mice carrying the inactivating *Met*[D/D] alleles have intercostal nerves that are much reduced in length and elaborate fewer terminal branches, showing a requirement for HGF/SF-*Met* activation in sensory neuron development (Maina et al., 1997). Interestingly, *Met* signaling via Grb2 is not required for sensory neuron development since *Met*[Grb2/Grb2] mice develop normally. Maina and colleagues also show that HGF cooperates with NGF in enhancing axonal growth of dorsal root ganglion and is required for survival of a proportion of neuron of dorsal root ganglion (Maina et al., 1997). If *Met* effectors different from Grb2 mediate HGF actions on sensory neurons, the same is not true for the responses elicited by this growth factor on sympathetic neurons. In fact, HGF- enhanced neurite outgrowth of NGF-dependent sympathetic neurons is not observed in mice carrying the *Met*[Grb2/Grb2] alleles (Maina et al., 1998). On the contrary, in these animals sympathetic neuroblast apoptosis is enhanced.

II. CONCLUDING REMARKS

The *Met* gene was originally identified as an oncogene (Cooper et al., 1984). Further studies showed that the tyrosine kinase receptor encoded by this gene plays a critical role during development. Activation of this receptor triggers a unique biological program leading to "invasive-cell-growth." This phenotype results from the integration of apparently independent biological responses such as cell proliferation, survival, motility, invasion of surrounding extracellular matrices and induction of cell polarity. During development, the coordinated control of "invasive growth" by the HGF/SF-*Met* pair is essential, as demonstrated by knock-out experiments involving either the ligand or the receptor. In adult life, deregulation of this "invasive-growth" program by the oncogenic forms of *Met* confers to cells transforming and invasive properties. Much is lacking for a complete understanding of the role played by *Met* in physiological and pathological processes, but it is tempting to speculate that further knowledge gained from the analysis of mice expressing *Met* signaling mutants will be critical to elucidate the signaling pathways responsible for "invasive growth."

ACKNOWLEDGMENT

Studies in the authors' laboratory were supported by grants from AIRC, Associazione Italiana per la Ricerca sul Cancro (Milano, Italy) and by The Giovanni Armenise-Harvard Foundation for Advanced Scientific Research (Cambridge, Massacheusetts). We thank Antonella Cignetto for secretarial assistance and Elaine Wright for help with the manuscript. We acknowledge Dr. Williams and Dr. Maina for the helpful discussion.

REFERENCES

Amicone L, Spagnoli F, Spät G, Giordano S, Tommasini C, Bernardini S, et al. (1996): Transgenic expression in the liver of truncated *Met* blocks apoptosis and permits immortalization of hepatocytes. EMBO J 16:495–503.

Andermarcher E, Surani M, Gherardi E (1996): Co-expression of the HGF/SF and c-*Met* genes during early mouse embryogenesis precedes reciprocal expression in adjacent tissues during organogenesis. Dev Genet 18:254–266.

Bardelli A, Basile ML, Audero A, Giordano S, Wennstrom S, Menard S, Comoglio PM, Ponzetto C (1999): Concomitant activation of pathways downstream of Grb2 and PI3 kinase is required for *Met*-mediated metastasis. Oncogene. In press.

Bladt F, Riethmacher D, Isenmann S, Aguzzi A, Birchmeier C (1995): Essential role for the c-*Met* receptor in the migration of myogenic precursor cells into the limb bud [see comments]. Nature 376:768–771.

Boccaccio C, Ando M, Tamagnone L, Bardelli A, Michieli P, Battistini C, et al. (1998): Induction of epithelial tubules by growth factor HGF depends on STAT pathway. Nature 391:285–288

Bottaro DP, Rubin JS, Faletto DL, Chan AM, Kmiecik TE, Vande Woude GF, et al. (1991): Identification of the hepatocyte growth factor receptor as the c-*Met* proto-oncogene product. Science 251:802–804.

Brand-Saberi B, Muller T, Wilting J, Christ B, Birchmeier C (1996): Scatter factor/ hepatocyte growth factor (SF/HGF) induces emigration of myogenic cells at interlimb level in vivo. Dev Bio 179:303–308.

Brinkmann V, Foroutan H, Sachs H, Weidner KM, Birchmeyer W (1995): Hepatocyte growth factor/scatter factor induces a variety of tissue specific morpgogenic programs in epithelial cells. J Cell Biol 131:1573–1586.

Bussolino F, Di Renzo MF, Ziche M, Bocchietto E, Olivero M, Naldini L, et al. (1992): Hepatocyte growth factor is a potent angiogenic factor which stimulates endothelial cell motility and growth. J Cell Biol 119:629–641.

Calvo E, Boucher C, Pelletier G, Morisset J (1996): Ontogeny of hepatocyte growth factor and c-*Met*/HGF receptor in rat pancreas. Bioc Biop R 4:257–263.

Cooper CS, Park M, Blair DG, Tainsky MA, Huebner K, Croce CM, et al. (1984): Molecular cloning of a new transforming gene from a chemically transformed human cell line. Nature 311:29–33.

Crepaldi T, Prat M, Giordano S, Medico E, Comoglio PM. (1994a): Generation of a truncated hepatocyte growth factor receptor in the endoplasmic reticulum. J Biol Chem 269:1750–1755.

Crepaldi T, Pollack AL, Prat M, Zborek A, Mostov K, Comoglio PM (1994b): Targeting of the SF/HGF receptor to the basolateral domain of polarized epithelial cells. J Cell Biol 125:313–320.

Detloff PJ, Lewis J, John SWM, Shehee WR, Langenbach R, Maeda N, Smithies O (1994): Deletion and replacement of the mouse adult beta-globin genes by a "plug and socket" repeated targeting strategy. Mol Cell B 14: 6936–6943

Donate LE, Gherardi E, Srinivasan N, Sowdhamini R, Aparicio S, Blundell TL (1994): Molecular evolution and domain structure of plasminogen-related growth factors (HGF/SF and HGF1/MSP). Protein Sci 3: 2378–2394

Ebens A, Brose K, Leonardo ED, Hanson MG Jr, Bladt F, Birchmeier C, Barres BA, Tessier-Lavigne M (1996): Hepatocyte growth factor/scatter factor is an axonal chemoattractant and a neurotrophic factor for spinal motor neurons. Neuron 17:1157–1172.

Ferracini R, Longati P, Naldini L, Vigna E, Comoglio PM (1991): Identification of the major autophosphorylation site of the *Met*/hepatocyte growth factor receptor tyrosine kinase. J Biol Chem 266:19558–19564.

Gandino L, Di Renzo MF, Giordano S, Bussolino F, Comoglio PM (1990): Protein kinase-c activation inhibits tyrosine phosphorylation of the c-*Met* protein. Oncogene 5:721–725.

Gandino L, Longati P, Medico E, Prat M, Comoglio PM (1994): Phosphorylation of serine 985 negatively regulates the hepatocyte growth factor receptor kinase. J Biol Chem 269:1815–1820.

Gherardi E, Gray J, Stoker M, Perryman M, Furlong R (1989): Purification of scatter factor, a fibroblast-derived basic protein that modulates epithelial interactions and movement. P NAS US 86:5844–5848.

Giordano S, Bardelli A, Zhen Z, Menard S, Ponzetto C, Comoglio PM (1997): A point mutation in the *Met* oncogene abrogates metastasis without affecting transformation P NAS US 94:13868–13782.

Giordano S, Di Renzo MF, Narsimhan RP, Cooper CS, Rosa C, Comoglio PM (1989b): Biosynthesis of the protein encoded by the c-*Met* proto-oncogene. Oncogene 4:1383–1388

Giordano S, Ponzetto C, Di Renzo MF, Cooper CS, Comoglio PM (1989a): Tyrosine kinase receptor indistinguishable from the c-*Met* protein. Nature 339:155–156.

Gohda E, Tsubouchi H, Nakayama H, Hirono S, Sakiyama O, Takahashi K, et al. (1988): Purification and partial characterization of hepatocyte growth factor from plasma of a patient with fulminant hepatic failure. J Clin Inv 81:414–419.

Gu H, Marth JD, Orban PC, Mossmann H, Rajewsky K (1994): Deletion of a DNA polymerase beta gene segment in T cells using cell type-specific gene targeting. Science 265:103–106.

Hanks M, Wurst W, Anson-Cartwright L, Auerbach AB, Joyner AL (1995): Rescue of the En-1 mutant phenotype by replacement of En-1 with En-2. Science 269:679–82.

Hasty P (1991): Introduction of a subtle mutation into the Hox-2.6 locus in embryopnic stem cells. Nature 350:243–246.

Hu Z, Evarts RP, Fujio K, Marsden ER, Thorgeirsson SS (1993): Expression of hepatocyte growth factor and c-*Met* genes during hepatic differentiation and liver development in the rat. Am J Path 142:1823–1830.

Jeffers M, Rao M, Rulong S, Reddy J, Subbarao V, Hudson E, et al. (1996): Hepatocyte growth factor scatter factor-*Met* signaling induces proliferation, migration, and morphogenesis of pancreatic oval cells. Cell Growth 7:1805–1813.

Kan M, Zhang GH, Zarnegar R, Michalopoulos G, Myoken Y, McKeehan WL, et al. (1991): Hepatocyte growth factor/hepatopoietin A stimulates the growth of rat kidney proximal tubule epithelial cells (RPTE), rat nonparenchymal liver cells, human melanoma cells, mouse keratinocytes and stimulates anchorage-independent growth of SV-40 transformed RPTE. Bioc Biop R 174:331–337.

Kawaguchi T, Qin L, Shimomura T, Kondo J, Matsumoto K, Denda K, et al. (1997): Purification and cloning of hepatocyte growth factor activator inhibitor type 2, a kunitz-type serine protease inhibitor. J Biol Chem 272: 27558–27564

Kobayashi T, Honke K, Miyazaki T, Matsumoto K, Nakamura T, Ishizuka I, et al. (1994) Hepatocyte growth factor specifically binds to sulfoglycolipids. J Biol Chem 269:9817–9821.

Kolatsijoannou M, Woolf A, Hardmann P, White S, Gordge M, Henderson R (1995): The hepatocyte growth factor scatter factor (HGF/SF) receptor, *Met*, transduces a morphogenic signal in renal glomerular fibromuscular mesangial cells. J Cell Sci 108:3703–3714.

Krasnoselsky A, Massay MJ, DeFrances MC, Michalopoulos G, Zarnegar R, Ratner N (1994): HGF is a mitogen for Scwann cells and is present in neurofibromas. J Neurosc 14:7284–7290.

Kunh R, Schwenk F, Aguet M, Rajewsky K (1995): Inducible gene targeting in mice. Science 269:1427–1429.

Lewis J, Yang B, Detloff P, Smithies O (1996): Gene modification via "Plug and Socket" gene targeting. J Clin Inv 1:3–5.

Liu Y, Suzuki K, Reed JD, Grinberg A, Westphal H, Hoffman A, Doring T, Sandhoff K, Proia RL (1998): Mice with type 2 and 3 Gaucher disease point mutations generated by a single insertion mutagenesis procedure (SIMP). P NAS US 95:2503–2508.

Longati P, Bardelli A, Ponzetto C, Naldini L, Comoglio PM (1994): Tyrosines 1234-1235 are critical for activation of the tyrosine kinase encoded by the *Met* proto-oncogene (HGF receptor). Oncogene 9:49–57.

Maina F, Casagranda F, Audero E, Simeone A, Comoglio PM, Klein, Ponzetto C (1996): Uncoupling of Grb2 from the *Met* receptor in vivo reveals complex roles in muscle development. Cell 87:531–542.

Maina F, Hilton MC, Ponzetto C, Davies AM, Klein R (1997): *Met* receptor signaling is required for sensory nerve development and HGF promotes axonal growth and survival of sensory neurons. Genes Dev 11:3341–3350.

Maina F, Hilton M, Andres R, Wyatt S, Klein R, Davies AM (1998): Multiple roles for HGF in sympathetic neuron development. Neuron 20:835–846

Mark MR, Lokker NA, Zioncheck TF, Luis EA, Godowski PJ (1992): Expression and characterization of hepatocyte growth factor receptor-IgG fusion proteins. Effects of mutations in the potential proteolytic cleavage site on processing and ligand binding. J Biol Chem 267:26166–26171.

Mars WM, Zarnegar R, Michalopoulos GK (1993): Activation of hepatocyte growth factor by the plasminogen activators uPA and tPA. Am J Path 143:949–958.

Matsumoto K, Tajima H, Nakamura T (1991): Hepatocyte growth factor is a potent stimulator of human melanocyte DNA synthesis and growth. Bioc Biop R 176:45–51.

Medico E, Mongiovì AM, Huff J, Jelinek MA, Follenzi A, Gaudino G, et al. (1996): The tyrosine kinase receptors Ron and Sea control "scattering" and morphogenesis of liver progenitor cells in vitro. Mol Biol Cell 7:495–504.

Miyazawa K, Shimomura T, Kitamura A, Kondo J, Morimoto Y, Kitamura N (1993): Molecular cloning and sequence analysis of the cDNA for a human serine protease responsible for activation of hepatocyte growth factor. Structural similarity of the protease precursor to blood coagulation factor XII. J Biol Chem 268:10024–10028

Miyazawa K, Tsubouchi H, Naka D, Takahashi K, Okigaki M, Arakaki N, et al. (1989): Molecular cloning and sequence analysis of cDNA for human hepatocyte growth factor. Bioc Biop R 163:967–973.

Mizuno K, Inoue H, Hagiya M, Shimizu S, Nose T, Shimohigashi Y, Nakamura T (1994): Hairpin loop and second kringle domain are essential sites for heparin binding and biological activity of hepatocyte growth factor. J Biol Chem 269: 1131–1136.

Montesano R, Matsumoto K, Nakamura T, Orci L (1991): Identification of a fibroblast-derived epithelial morphogen as hepatocyte growth factor. Cell 67:901–908

Nagaike M, Hirao S, Tajima H, Noji S, Taniguchi S, Matsumoto K, et al. (1991): Renotropic functions of hepatocyte growth factor in renal regeneration after unilateral nephrectomy. J Biol Chem 266:22781–22784.

Nakamura T, Nishizawa T, Hagiya M, Seki T, Shimonishi M, Sugimura A, et al. (1989): Molecular cloning and expression of human hepatocyte growth factor. Nature 342:440–443.

Nakamura T, Teramoto H, Ichihara A (1986): Purification and characterization of a growth factor from rat platelets for mature parenchymal hepatocytes in primary cultures. P NAS US 83:6489–6493.

Naldini L, Tamagnone L, Vigna E, Sachs M, Hartmann G, Birchmeier, et al. (1992): Extracellular proteolytic cleavage by urokinase is required for activation of hepatocyte growth factor/scatter factor. EMBO J 11:4825–4833.

Naldini L, Vigna E, Ferracini R, Longati P, Gandino L, Prat M, et al. (1991): The tyrosine kinase encoded by the *Met* proto-oncogene is activated by autophosphorylation. Mol Cell B 11:1793–1803.

Naldini L, Vigna E, Narsimhan RP, Gaudino G, Zarnegar R, Michalopoulos GK, et al. (1991a): Hepatocyte growth factor (HGF) stimulates the tyrosine kinase activity of the receptor encoded by the proto-oncogene c-*Met*. Oncogene 6:501–504.

Naldini L, Weidner KM, Vigna E, Gaudino G, Bardelli A, Ponzetto C, et al. (1991): Scatter factor and hepatocyte growth factor are indistinguishable ligands for the *Met* receptor. EMBO J 10:2867–2878.

Pelicci G, Giordano S, Zhen Z, Salcini AE, Lanfrancone L, Bardelli A, et al. (1995): The motogenic and mitogenic responses to HGF are amplified by the Shc adaptor protein. Oncogene 10:1631–1638.

Pepper MS, Soriano JV, Menoud PA, Sappino AP, Orci L, Montesano R (1995): Modulation of hepatocyte growth factor and c-*Met* in the rat mammary gland during pregnancy, lactation, and involution. Exp Cell Re 219:204–210.

Ponzetto C, Bardelli A, Zhen Z, Maina F, dalla Zonca P, Giordano S, et al. (1994): A multifunctional docking site mediates signaling and transformation by the hepatocyte growth factor/scatter factor receptor family. Cell 77:261–271.

Ponzetto P, Zhen Z, Audero E, Maina F, Bardelli A, Basile M, et al. (1996): Specific uncoupling of GRB2 from the *Met* receptor. J Biol Chem 271:14119–14123.

Ridley AJ, Comoglio PM, Hall A (1995): Regulation of scatter factor/hepatocyte growth factor responses by Ras, Rac, and Rho in MDCK cells. Mol Cell B 15: 1110–1122.

Royal I, Park M (1995): Hepatocyte Growth Factor-induced scatter of Madin-Darby canine kidney cells requires phosphatidylinositol 3-kinase. J Biol Chem 270:27780–27787.

Sachs M, Weidner KM, Brinkmann V, Walther I, Obermeier A, Ullrich A, et al. (1996): Motogenic and morphogenic activity of epithelial receptor tyrosine kinases. J Cell Biol 133:1095–1107.

Santos OFP, Barros EJG, Yang XM, Matsumoto K, Park M, Nigam SK (1994): Involvment of HGF in kidney development Dev Biol 163:525–529.

Schmidt C, Bladt F, Goedecke S, Brinkmann V, Zschiesche W, Sharpe M, et al. (1995): Scatter factor/hepatocyte growth factor is essential for liver development. Nature 373:699–702.

Shimomura T, Miyazawa K, Komiyama Y, Hiraoka H, Naka D, Morimoto Y, et al. (1995): Activation of hepatocyte growth factor by two homologous proteases, blood-coagulation factor XIIa and hepatocyte growth factor activator. Eur J Bioch 229:257–261.

Shiota G, Wang T, Nakamura T, Schmidt E (1994): Hepatocyte growth factor in transgenic mice: effects on hepatocyte growth, liver regeneration and gene expression. Hepatology 19:962–972.

Sonnenberg E, Meyer D, Weidner KM, Birchmeier C (1993): Scatter factor/hepatocyte growth factor and its receptor, the c-*Met* tyrosine kinase, can mediate a signal exchange between mesenchyme and epithelia during mouse development. J Cell Biol 123:223–235.

Stern CD, Ireland GW, Herrick SE, Gherardi E, Gray J, Perryman M, et al. (1990): Epithelial scatter factor and development of the chick embryonic axis. Development 110:1271–1284.

Stoker M, Gherardi E, Perryman M, Gray J (1987): Scatter factor is a fibroblast-derived modulator of epithelial cell mobility. Nature 327:239–242.

Tabata M, Kim K, Liu J, Yamashita K, Matsumura T, Kato J, et al. (1996): Hepatocyte growth factor is involved in the morphogenesis of tooth germ in murine molars. Development 122:1243–1251.

Takayama H, LaRochelle W, Anver M, Bockman D, Merlino G (1996): Scatter factor/hepatocyte growth factor as a regulator of skeletal muscle and neural crest development. P NAS US 93:5866–5871.

Thery C, Sharpe MJ, Batley SJ, Stern CD, Gherardi E (1995): Expression of HGF/SF, HGF1/MSP, and c-*Me*t suggests new functions during early chick development. Dev Genet 17:90–101.

Villa-Moruzzi E, Lapi S, Prat M, Gaudino G, Comoglio PM (1993): A protein tyrosine phosphatase activity associated with the hepatocyte growth factor/scatter factor receptor. J Biol Chem 268:18176–18180.

Weidner KM, Behrens J, Vandekerckhove J, Birchmeier W (1990): Scatter factor: molecular characteristics and effect on the invasiveness of epithelial cells. J Cell Biol 111:2097–2108.

Weidner KM, Arakaki N, Hartmann G, Vandekerckhove J, Weingart S, Rieder H, et al. (1991): Evidence for the identity of human scatter factor and human hepatocyte growth factor. P NAC US 88:7001–7005.

Weidner K, DiCesare S, Sachs M, Brinkmann V, Behrens J, Birchmeier W (1996): Interaction between Gab1 and the c-*Met* receptor tyrosine kinase is responsible for epithelial morphogenesis. Nature 384:173–176.

Weidner KM, Sachs M, Riethmacher D, Birchmeier W (1995): Mutation of juxtamembrane tyrosine residue 1001 suppresses loss of function mutations of the *Met* receptor in epithelial cells. P NAS US 92:2597–2601.

Yamagata T, Muroya K, Mukasa T, Igarashi H, Momoi M, Tsukahara T, et al. (1995): HGF specifically expressed in microglia acivated Ras in the neurons, similar to the action of neurotrophic factros. Bioc Biop R 210:231–237.

Yanagita K, Matsumoto K, Sekiguchi K, Ishibashi H, Niho Y, Nakamura T (1993): Hepatocyte growth factor may act as a pulmotrophic factor on lung regeneration after acute lung injury. J Biol Chem 268: 21212–21217.

Yang Y, Park M (1993): Expression of the *Met*HGF/SF receptor and its ligand during differentiation of murine P19 embryonal carcinoma cells. Dev Biol 157:308–320.

Yang Y, Spitzer E, Meyer D, Sachs M, Niemann C, Hartmann G, et al. (1995): Sequential requirement of hepatocyte growth factor and neuregulin in the morphogenesis and differentiation of the mammary gland. J Cell Biol 131:215–226.

Yang X, Vogan K, Gros P, Park M (1996): Expression of the *Met* receptor tyrosine kinase in muscle progenitor cells in somites and limbs is absent in Splotch mice. Development 122:2163–2171.

Zarnegar R, Michalopoulos G (1989): Purification and biological characterization of human hepatopoietin A, a polypeptide growth factor for hepatocytes. Cancer Res 49:3314–3320.

CHAPTER 8

TISSUE-SPECIFIC KNOCKOUTS OF THE INSULIN RECEPTOR TYROSINE KINASE

ROHIT N. KULKARNI, M.D., PH.D;
M. DODSON MICHAEL, PH.D., and C. RONALD KAHN, M.D.

I. THE TYROSINE KINASE RECEPTOR FAMILY

Development and survival of multicellular organisms is dependent on a coordinated communication between cells. Cell-cell communication is established through polypeptides, including hormones and growth and differentiation factors. Many of these factors mediate their actions via binding to cell-surface receptors that possess intrinsic tyrosine kinase activity. The tyrosine kinase receptors generally have a similar configuration—an extracellular ligand-binding domain, a hydrophobic transmembrane region, and a cytoplasmic domain with tyrosine kinase catalytic activity (Schlessinger, 1988; Hanks et al., 1988).

Based on structural configuration, the different cell-surface receptors with tyrosine kinase properties have been classified into four subclasses. Subclass I receptors are monomeric and feature two cysteine-rich repeat sequences, whereas subclass II receptors also contain cysteine-rich sequences but are heterotetrameric ($\alpha_2\beta_2$) structures. The presence of immunoglobulin-like repeats characterizes subclass III (five repeats) and subclass IV (three repeats) categories, the latter with varying lengths of hydrophilic insertion sequences. Examples of the various factors that belong to these different classes are shown in Figure 8.1a. The insulin receptor is representative of subclass II.

2. THE INSULIN RECEPTOR

The insulin receptor has an apparent molecular weight (Mr) of 350,000 and is composed of two α and two β subunits covalently linked through disulfide bonds to form an $\alpha_2\beta_2$ heterotetramer (Fig. 8.1b). With reduction of SDS-PAGE, the α

Genetic Manipulation of Receptor Expression and Function,
Edited by Domenico Accili.
ISBN 0-471-35057-5   Copyright © 2000 Wiley-Liss, Inc.

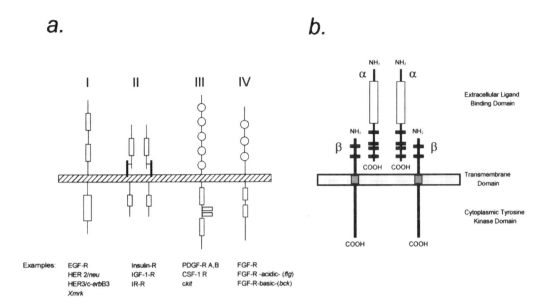

Figure 8.1. The Tyrosine Kinase family of Receptors. (a) The tyrosine kinase receptors are classified based on structural configuration into subclass I and II that contain cysteine-rich sequences (shown as rectangles) and subclass II and IV that contain immunoglobulin-like repeats (shown as open circles). Examples of receptors belonging to the different subclasses is shown below the respective structure. (b) The insulin receptor is composed of two α and two β subunits that constitute an extracellular ligand-binding domain, a transmembrane domain, and a cytoplasmic tyrosine kinase domain. Potential tyrosine phosphorylation sites are denoted by black boxes.

and β subunits have an estimated Mr of 135,000 and 95,000, respectively (Czech, 1981; Kahn and White, 1988). The α subunit contains the insulin-binding domain, and the β subunit possesses a protein tyrosine kinase-domain with each subunit performing specific functions in signal transduction. Each subunit possesses several cysteine residues that link the subunits to form the heterotetramer.

The insulin receptor gene is located close to the low-density lipoprotein receptor gene on the short arm of chromosome 19 (Gammeltoft et al., 1985). The gene (150 kb in length) contains 22 exons separated by long introns. The cDNA encoding the human insulin proreceptor was isolated and analyzed, revealing a site of alternative splicing surrounding exon 11 (Ullrich et al., 1985). This encoding results in two alternatively spliced insulin receptors, each arising from a distinct mRNA species that encodes the subunits (Seino and Bell, 1989). Some of the heterogeneity that has been reported in insulin-binding among different tissues can be partially explained by the differential expression of these two receptor types in different tissues.

3. THE INSULIN-SIGNALING CASCADE

The insulin receptor mediates the actions of insulin—one of the most important hormones required for normal physiological functioning in mammals. The mechanism of insulin action following receptor-ligand interaction is complex and has been the subject of intensive investigation over the past several decades at both the whole animal level and at the cellular level. Insulin elicits several physiological responses, including stimulation of glucose uptake; lipid, glycogen, and protein synthesis; and inhibition of lipolysis in key peripheral tissues such as liver, skeletal muscle, and adipose. However, the precise biochemical pathways that generate each of these processes and how they are linked to the receptor are still only partly understood. An understanding of these molecular processes is essential to unravel the pathogenesis underlying several diseases manifesting insulin resistance, especially non-insulin dependent diabetes (type 2 diabetes) and obesity.

4. APPROACHES TO THE STUDY OF THE INSULIN-SIGNALING CASCADE

In Vitro Systems

Several approaches have been used to study the molecular action of insulin following binding of the hormone to the receptor. The availability of cDNA clones has made it possible to analyze the structure-function relationship of the insulin receptor in considerable detail. Studies using over-expression systems and dominant negative mutants in vitro have revealed some of the early events following the binding of insulin to the insulin receptor. Among the early signaling events in the insulin-signaling cascade—common to all tyrosine kinase signaling mechanisms—is ligand-induced dimerization of the receptor followed by downstream protein-protein interactions initiated by the activated receptor (reviewed in Cheatham and Kahn, 1995). Studies employing in vitro mutagenesis have defined some of the α subunit residues of the insulin receptor involved in insulin-binding (Cheatham and Kahn, 1992) and have been useful in identifying the residues involved in the disulfide linkage between the α and β subunits. Similarly a series of studies have defined the role of the transmembrane domain of the insulin receptor β subunit in regulating kinase activation and receptor function (Krempler et al., 1998; Goncalves et al., 1993; Cheatham et al., 1993). Specificity of the signaling mechanism begins with autophosphorylation of multiple tyrosine residues and selective binding of specific SH2-proteins to the phosphorylated tyrosines. The insulin receptor has been shown to initiate a signaling cascade by phosphorylation of insulin substrate (IRS) proteins, Shc and Gab-1. Phosphorylated IRS-1 and IRS-2 bind to the regulatory subunit of PI 3-kinase and activate PI 3-kinase during its course of binding to SH2 proteins (Cheatham et al., 1994). In addition several other SH2 proteins, including GRB2, SHIP, SHP-2 and nck, have been shown to be associated with the IRS family of proteins in the course of producing the pleiotropic effects of insulin. Although several proteins and regulatory steps in the signaling pathway have been defined in vitro, the steps linking these events to the final effects of insulin action, such as protein and DNA synthesis, continue to be elusive.

Studies of naturally occurring mutants of the molecules in the insulin-signaling cascade have been useful to some extent in understanding the role of specific residues in the insulin receptor and various substrates of the insulin receptor. Mutations of the insulin receptor have been reported to cause syndromes associated with severe insulin resistance, including leprechaunism and type A insulin resistance (Taylor, 1992; Flier et al., 1975). One the one hand, amino-acid sequence variants in IRS-1 have been identified at a higher proportion in NIDDM patients compared with the normal population (Laakso et al., 1994; Almind et al., 1993; Imai et al., 1994). On the other hand, diabetes does not appear to be associated with mutations or polymorphisms in other molecules of the insulin-signaling system such as IRS-2 (Withers et al., 1998) and the p85 subunit of PI 3-kinase (Terauchi et al., 1999). These mutations, however, may account for only a subset of the patients with type 2 diabetes.

Whole-Body Gene Targeting

One recent approach to determine the role of the multiple insulin-signaling proteins in insulin action has been the disruption of genes that encode these proteins. We and others have produced transgenic and knock-out models of NIDDM by deleting single or multiple genes, yet none of these models have been designed to mimic the early skeletal muscle insulin resistance present in individuals destined to develop NIDDM.

Mice with a whole-body knock-out of the IR gene die within a few days after birth from severe ketoacidosis and metabolic abnormalities. This result suggests that insulin mediates most of its metabolic effects via the insulin receptor, however, early death in these mice prevents a detailed analysis of insulin receptor function in individual tissues (Accili et al., 1996; Joshi et al., 1996; Joshi et al., 1996). By contrast, knockout of the IGF1 receptor results in severe intrauterine growth retardation and death due to respiratory insufficiency. In addition, several genes coding for proteins in the insulin-signaling cascade have been knocked out in an effort to study the consequent effects on glucose metabolism. Inactivation of IRS-1 (Araki et al., 1994; Tamemoto et al., 1994), a primary substrate for the insulin and IGF receptors, results in a milder defect with 50% growth retardation and insulin resistance. In addition, these mice have an insulin secretory defect (Kulkarni et al. JCI, 1999) and females are infertile (Kulkarni RN and Kahn CR, unpublished data). These data suggest that the insulin-signaling pathway likely has multiple components with alternate substrates taking over the signaling of the IRS-1 molecule. Mice with a knockout of the IRS-2 (Withers et al., 1998), the p85α subunit of PI 3-kinase (Terauchi et al., 1999), or Glut4 (Katz et al., 1995) genes all exhibit varying disturbances in glucose homeostasis, but none provides a clear "muscle-specific" defect that might mimic early NIDDM. Combined heterozygous defects in the insulin receptor and IRS-1 results in a more severe phenotype with mice showing hyperinsulinemia early in life followed by severe insulin resistance and massive islet hyperplasia by 6 months of age (Bruning et al., 1997). It is notable that in spite of the islet hyperplasia, ~50% of the mice develop diabetes. Thus combined heterozygous defects in two molecules in the insulin-signaling system demonstrate the role of epistatic interactions in the pathogenesis of common diseases with non-Mendelian genetics.

An alternate attempt to specifically address the role of insulin receptor signaling in skeletal muscle in glucose homeostasis used mice expressing a dominant negative mutant of the insulin receptor under control of the MCK promoter (Chang et al., 1994; Moller et al., 1996). It is difficult, however, to interpret these data because the effect of decreased insulin-receptor tyrosine kinase activity on insulin-stimulated glucose transport in isolated muscles was variable, and glucose uptake in isolated soleus muscle of the insulin receptor dominant negative mice was not affected (Moller et al., 1996). Recently Lauro et al. produced transgenic mice expressing a kinase-defective human insulin receptor under the control of the human insulin receptor promoter (Lauro et al., 1998). Although these mutant mice developed several features observed in early diabetes in humans, none of the mice became overtly diabetic. One caveat to the interpretation of these dominant negative approaches is that insulin receptors are known to dimerize with IGF-1 receptors in skeletal muscle, and the effects of these mutant heterodimers on insulin and IGF-1 signaling were not addressed. Thus, none of the previous approaches has been effective in clarifying the exact contribution of insulin signaling in individual tissues to whole-body glucose metabolism and the development of NIDDM. In these studies, the different phenotypes observed, ranging from mild defects to severe diabetes with ketoacidosis and early death, suggest the existence of complementarity between key molecules in the signaling pathway.

5. CONDITIONAL GENE TARGETING WITH THE CRE/*LOXP* SYSTEM

Although whole-body null mutations give some idea of the function of the insulin receptor, the significance of insulin signaling in specific tissues known to play a role in type 2 diabetes may be better addressed by the generation of tissue-specific knock-outs. Recently, two approaches using site-specific DNA recombinases have been developed for conditional gene targeting. The most widely used of these is Cre recombinase, a 38-kDa site-specific DNA recombinase from the bacteriophage P1 (Sternberg et al., 1986). Cre recognizes a 34-bp DNA element, called *loxP*, which consists of two 13-bp inverted repeats flanking an 8-bp central region (Fig. 8.2) (Hoess et al., 1990). The non-palindromic nature of the central core of the *loxP* site gives it directionality (Fig. 8.2). Recombination between two directly repeated *loxP* sites results in the excision of the intervening DNA as a closed loop, leaving one of the *loxP* sites in the original target sequence (Fig. 8.2, left panel). However, recombination between two *loxP* sites that point towards each other results in the inversion of the intervening sequences (Fig. 8.2, right panel).

In 1994, Rajewsky and coworkers introduced a technique to conditionally inactivate genes in mice by using Cre-*loxP*-mediated recombination (Gu et al., 1994). This strategy for generating spatially or temporally regulated gene knockouts requires two steps (Fig. 8.3). First, the endogenous gene to be targeted is modified by homologous recombination in murine embryonic stem (ES) cells so that *loxP* sites flank a portion of the gene. Mice are generated from these recombinant ES cells by standard techniques to give rise to a strain of "flox" (flanked by lox) mice. Second, the heterozygous flox mice are bred with mice that express a *cre* transgene to generate double heterozygous mice

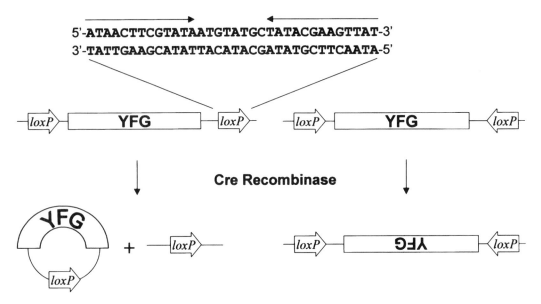

Figure 8.2. Cre/*loxP* DNA recombination. The *loxP* site consists of two 13-bp inverted repeats (marked with thin arrows) flanking an 8-bp non-palindromic central region, which gives the site directionality (depicted as thick arrow). Recombination between two directly repeated *loxP* sites results in the excision of the intervening DNA as a closed loop leaving one of the *loxP* sites in the original target sequence (left pathway). Recombination between two *loxP* sites that point towards each other results in the inversion of the intervening sequences (right pathway). *YFG*, your favorite gene.

[Fig. 8.3, YFG(*lox*/+):Cre(+/)]. These double heterozygous mice are bred with single heterozygous flox mice, giving rise to the conditional knock-out mice [Fig. 8.3, YFG(*lox/lox*):Cre(+/)] as well as several control groups. Cre-mediated recombination between the engineered *loxP* sites results in gene disruption in vivo in only the tissue(s) where Cre is expressed. One of the most powerful and attractive features of this system is that a single line of floxed mice can be used for gene disruption in many tissues or at different times in development depending on the promoter that is used to drive expression of the *cre* transgene. An international collaboration has led to the development of a database of transgenic mice expressing Cre recombinase with different spatial and temporal specificities (http://www.mshri.on.ca/develop/nagy/Cre.htm).

Creation of Mice Carrying *loxP*-flanked Alleles

Both steps of the conditional gene targeting require careful planning and precise characterization to ensure the success of the knockout. The positioning of the *loxP* sites and the overall design of the targeting vector are two of the most crucial elements in generating an efficacious flox mouse. Because the *loxP* site has a specific 34-bp sequence, most insertions of *loxP* sites are in introns or in the 5′- or 3′-untranslated portions of the target gene to prevent disruptions of the coding sequence.However, one must try to avoid any regulatory elements or splicing domains, because these factors could result in altered gene expression even in the absence of Cre.

For *loxP* targeting of the IR (*Insr*) gene (Bruning et al., 1998), we created a targeting vector that contained a *neo*-HSV-*tk* drug-selection cassette flanked by directly repeated *loxP* sites upstream of exon 4 of the IR gene and a single *loxP* site 80 bp downstream of exon 4 (Fig. 8.4a). To introduce a negative selection cassette for non-homologous recombination events, the diphtheria toxin-A (DTA) cDNA driven by the thymidine kinase promoter was positioned upstream of the 5'-region of homology. Exon 4 of the insulin receptor gene was targeted for Cre-mediated deletion because the splicing of exon 3 to exon 5 causes a frameshift mutation resulting in a stop of translation at amino acid 308

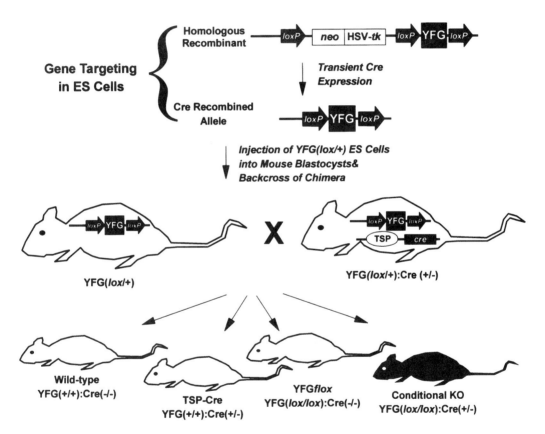

Figure 8.3. Generation of conditional knockout mice by Cre/*loxP* techniques. A targeting vector that contains your favorite gene (*YFG*) flanked by *loxP* sites and *loxP*-flanked selectable marker (*neo/HSV-tk*) are transfected in murine ES cells and homologous recombinants are selected. Transient Cre expression in the ES cells is used to remove the selection cassette, and the gene targeted ES cells are injected into murine blastocysts to generate chimeric mice. Chimeric mice are backcrossed and characterized according to standard procedures. Mating of a heterozygous flox mouse [YFG(*lox*/+)] with a heterozygous flox mouse that carries a Cre transgene [YFG(*lox*/+):Cre(+/−)] under the control of a tissue-specific promoter (*TSP*) gives rise to the conditional knockout mice [YFG(*lox/lox*):Cre(+/−)] as well as several control groups.

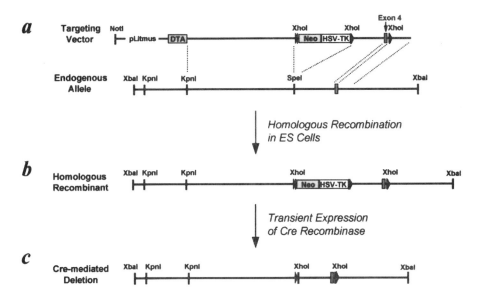

Figure 8.4. *LoxP* targeting of the insulin receptor gene. (a) The upper diagram shows the targeting vector that was constructed to introduce *loxP* sites into the insulin receptor gene. A *loxP*-flanked *neo*/HSV-*tk* selection cassette was placed into intron 3 ~2 kb upstream of exon 4. A third *loxP* site was placed 80 bp downstream of exon 4. The diphtheria toxin A (*DTA*) cDNA was placed outside of the region of homology as a negative selection marker for non-homologous recombination events. The lower diagram shows a simplified restriction map of the endogenous murine insulin receptor locus surrounding exon 4 (*open box*). (b) Following transfection of the targeting vector into murine ES cells, clones in which homologous recombination had occurred were selected by growth in G418-containing media. Homologous recombination was confirmed by Southern blotting. (c) Cre-recombinase was transiently expressed in the recombinant ES cells to delete the drug-selection cassette. Excisional recombination of the selection cassette was selected by growth in ganciclovir-containing media. Because this selection cannot distinguish clones that have undergone complete recombination from those that have only lost the drug selection cassette, appropriate recombination was confirmed by Southern blot and PCR analysis.

thus producing a null allele. It must also be noted that positioning of the *loxP* sites around other exons, such as exon 2, could produce in-frame deletions that may not affect the function of the protein produced; therefore, careful planning of the positions of the *loxP* sites is important. For smaller, less-complex genes, it may be possible to flank the entire coding region of the target gene with *loxP* sites. The targeting construct is transfected into ES cells by using standard techniques, and homologous recombinants (Fig. 8.4b) are positively selected by G418-resistance and confirmed by Southern blotting.

Because the *neo* gene that is used as a positively selectable marker could potentially interfere with target gene expression, we believe that it is advisable to remove the *neo* cassette. The triple *loxP* strategy described for the IR targeting vector (Fig. 8.4b) provides two *loxP*-flanked intervals—exon 4 of the IR gene and the selection cassette—that can be removed by Cre-mediated recombination. Complete recombination would remove both of the *loxP*-flanked intervals; however, transient Cre expression in the targeted ES cells can lead to some clones of cells in which partial recombination has occurred, resulting in deletion of only the selection cassette (Fig. 8.4c). The HSV-*tk* gene can be included in the *loxP*-flanked drug-selection cassette to allow for selection of ganciclovir-resistant clones that have lost the drug-selection cassette. Because this selection cannot distinguish clones that have undergone complete recombination from those that have lost only the drug-selection cassette, Southern blot and PCR analysis must be used to demonstrate the presence of the *loxP*-flanked target gene region and the loss of the selection cassette. Correctly recombined clones have single *loxP* sites arranged as direct repeats upstream and downstream of the targeted region (Fig. 8.4c). An alternative approach to the *loxP*-flanked selection cassette is to flank the selection cassette with FRT sites—recognition sites for the yeast DNA recombinase, Flp (Andrews et al., 1985). The Flp/FRT system has been used for site-specific excisional recombination in murine ES cells as well as in transgenic mice (Fiering et al., 1995; Dymecki, 1996). The advantage of the Flp/FRT system is that only one recombination event is possible, which is removal of the selection cassette. Once the selection cassette is excised, the heterozygous flox ES cell clones are injected into blastocysts, and chimeric mice carrying the flox allele are characterized according to standard procedures (Papaioannou and Johnson, 1993).

A major concern for in vitro excision of the drug-selection cassette is that multiple manipulations of ES cells may result in decreased ability of the ES cells to contribute to the germline in chimeric mice. For removal of a simple *loxP*-flanked *neo* positive selection cassette, the second in vitro ES cell selection can be avoided by Cre-mediated recombination in the mouse to remove the selectable marker. In brief, a targeting vector is constructed containing a triple *loxP*-flanked target gene, including a *loxP*-flanked *neo* selection cassette and an HSV-*tk* gene outside of the region of homology for negative selection. The ES cells are electroporated with this construct and selected by G418 and ganciclovir. Recombinant ES cell clones harboring the triple *loxP*-flanked target gene are injected into blastocysts to generate chimeric mice. The chimera is then either mated with a Cre-expressing mouse, such as CMV-*cre* (Lakso et al., 1992; Schwenk et al., 1995) or EIIa-*cre* (Lakso et al., 1996), or zygotes from a chimera × wild-type (WT) mating can be microinjected with a Cre expression plasmid or *cre* mRNA (Araki et al., 1995; Sunaga et al., 1997). Any of these strategies will promote Cre-mediated recombination between the *loxP* sites, but because there are three potential deletion events, Southern blot and/or PCR analysis must be performed to identify offspring carrying the appropriate deletion event. Because the EIIa promoter is expressed only from the one-cell zygote stage until the time of implantation (Dooley et al., 1989), the EIIa-*cre* mouse may be the method of choice for in vivo deletion of *loxP*-flanked *neo* selection cassettes.

Design and Analysis of *cre* Transgenic Mice

The second crucial factor in generating efficient and interpretable conditional gene-targeted mice is the design and analysis of the *cre* transgenic mouse. The initial descriptions of the Cre/*loxP* system noted inefficient (<50%) deletion of the targeted gene (Gu et al., 1994); however, modifications to the *cre* cDNA to include a consensus Kozak sequence has alleviated one problem with *cre* expression in mammalian cells (Kozak, 1999; Sauer et al., 1990). Because the prokaryotic Cre protein might not be efficiently transported into eukaryotic nuclei, Rajewsky and colleagues as well as others engineered the SV40 nuclear localization signal (NLS) into *cre* transgenes (Gu et al., 1994; Bruning et al., 1998). The necessity of this modification has never been directly tested; however, nuclear targeting of a GFP-Cre fusion protein that lacked an exogenous NLS may support the argument that the NLS modification is not essential (Gagneten et al., 1997).

The selection of a well-characterized promoter to drive expression of the *cre* transgene with the desired spatial and temporal specificity is a prerequisite for the generation of a *cre* transgenic mouse. Unexpected expression of Cre in tissues other than the intended target tissue at any time during development complicates the interpretation of any resultant phenotype because even brief Cre expression can lead to complete gene deletion in the cell where Cre was expressed as well as in any daughter cells. Unfortunately, even careful selection of a promoter cannot eliminate mosaic or partial Cre expression in the desired tissue. Mosaic transgene expression can result from the random site of transgene integration as well as from position effects of the promoter (Dobie et al., 1996). In some instances, the *cre* gene has been knocked into an endogenous gene locus with a known pattern of expression to circumvent the problems of mosaicism (Chen et al., 1998).

To ensure that the spatial and temporal specificity of Cre expression is as desired, a full characterization of all novel *cre* transgenic mice should be performed. RT-PCR and Northern blot analysis have been used to screen for the expression of *cre* mRNA, and antibodies have become available that allow for Western and immunohistochemical detection of Cre expression in mouse tissues. In addition, several laboratories have generated strains of transgenic mice that express *loxP*-disrupted reporter genes from a ubiquitously expressed promoter (Lobe et al., 1999; Mao et al., 1999). Crossing these reporter mice with *cre* transgenic mice followed by histological analysis of reporter gene expression in the offspring provides a powerful analysis of Cre function in the whole animal.

For the production of muscle-specific Cre transgenic mice, we chose a 6.5-kb genomic DNA fragment from the muscle creatine kinase (MCK) gene for several reasons. First, previous studies have demonstrated that expression of MCK is highly restricted to skeletal muscle and heart (Trask and Billadello, 1990). Second, it has been shown that this 6.5-kb fragment containing the MCK promoter, enhancer 1, the untranslated exon 1 and intron 2 containing enhancer 2 can confer this regional specificity to a downstream coding sequence under transgenic conditions (Johnson et al., 1989). In addition, expression of MCK begins at embryonic day 17 in the rat, increases to ~40% of maximal levels at birth, reaches maximal levels at day 10, and remains at a constantly high level throughout the rest of life (Trask and Billadello, 1990). This pattern of expression of sustained Cre expression may help overcome mosaic transgene expression, especially due to the syncytial nature of muscle fibers.

The Cre cDNA containing a nuclear localization signal was subcloned in place of the translation initiation site of the MCK gene, followed by an SV40 polyadenylation signal. Injection of the transgene resulted in the creation of 12 independent lines of mice, of which seven transmitted the transgene through their germline. For the creation of the tissue-specific knockout, we chose the mouse that had the highest level of Cre in skeletal muscle. Analysis of RNA extracted from different tissues of this line revealed high levels of Cre mRNA in skeletal muscle, weak expression in cardiac muscle, and no detectable mRNA in a variety of other tissues. Restriction of Cre protein expression to the muscle of this line was further verified by Western blot analysis.

6. CREATION OF MUSCLE-SPECIFIC INSULIN RECEPTOR KNOCKOUT MICE

Breeding of IR(*lox*/+) and MCK-Cre(+/−) mice resulted in double heterozygous animals that were then bred with IR(*lox*/+) mice to obtain IR(*lox/lox*):MCK-Cre offspring, that is mice with a muscle-specific insulin receptor knockout (MIRKO). To investigate the effect of the MCK-Cre transgene on the IR*lox* allele, protein extracts from hind-limb muscles were subjected to immunoprecipitation with IR-specific antisera followed by Western blot analysis using the same antibody. As shown in Figure 8.5a, insulin receptor expression was unaltered in muscle from the control IR(*lox/lox*) mice as compared with that from WT animals, indicating that the introduction of the *loxP* sites in the insulin receptor allele did not interfere with IR expression. By contrast, IR expression was reduced by 95% in pooled hind-limb muscle from the MIRKO mice (Fig. 8.5a). IR expression was reduced by 90–99% in quadriceps, gastrocnemius, gluteus, soleus, EDL, triceps, and pectoralis and even reduced by 92% in heart (which had considerably lower Cre expression) when extracts from MIRKO mice were compared with those of IR(*lox/lox*) and WT mice (Fig. 8.5b). Because skeletal muscle samples contain an admixture of other tissues, we performed immunohistochemical analysis on tissue sections prepared from mice of the different genotypes to determine the origin of the faint residual immunoreactive IR protein in some muscle samples of the MIRKO animals. Whereas WT and IR(*lox/lox*) mice exhibited homogenous staining of myocytes as well as adjacent adipose tissue and vascular endothelial cells "contaminating" the sample, there was no detectable signal in muscle of the MIRKO mice despite comparable staining of other cell types (Fig. 8.5c).However, Western blot analysis revealed that the levels of IR protein in non-muscle tissues from the MIRKO mice were not different from controls (Fig. 8.5d). Thus, expression of the Cre transgene under control of the MCK promoter was sufficient to direct recombination between the *loxP* sites with high efficiency in muscle, independent of anatomical localization or fiber types, and was highly effective in abolishing IR expression in all skeletal muscles and in heart.

Characterization of the MIRKO Mouse

To analyze the consequence of muscle-specific knockout of IR on insulin-stimulated signaling on a molecular level, we compared the early steps of insulin signaling in skeletal muscle and liver of the MIRKO mice. Whereas

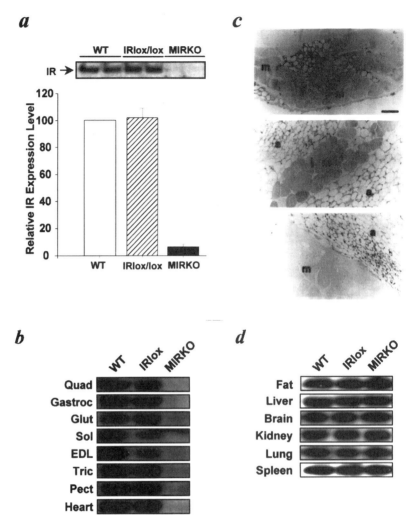

Figure 8.5. MIRKO mice exhibit high-efficiency inactivation of the insulin receptor gene selectively in skeletal muscle and heart. (a) Immunoprecipitation and Western blot analysis on hind-limb muscle extracts from WT (*open bar*), IR(*lox/lox*) (*hatched bar*), and MIRKO (*filled bar*) mice using an insulin receptor-specific antiserum. The graph shows the quantification of insulin receptor content as percent of insulin receptor content in muscle from WT mice. (b) Protein extracts from different muscles of WT, IR(*lox/lox*), and MIRKO mice were subjected to immunoprecipitation with an insulin receptor-specific antiserum followed by Western blot analysis using the same antibody. *Quad*, quadriceps; *Gastroc*, gatrocnemius; *Glut*, gluteus; *Sol*, soleus; *EDL*, extensor digitorum longus; *Tric*, triceps; *Pect*, pectoralis. (c) Immunohistochemical analysis of muscle isolated from WT (upper panel), IR(*lox/lox*) (middle panel), and MIRKO (lower panel) mice using an insulin receptor-specific antiserum. Note the lack of insulin receptor expression in muscle fibers (*m*) from MIRKO mice, whereas contaminating adipose tissue (*a*) exhibits staining equal to WT and IR(*lox/lox*) mice. (d) Protein extracts were prepared from tissues of WT, IR(*lox/lox*), and MIRKO mice and subjected to immunoprecipitation with an insulin receptor-specific antiserum followed by Western blot analysis using the same antibody.

insulin signaling in liver was unaltered, it was virtually abolished in skeletal muscle of the MIRKO mice (Fig. 8.6a). Similarly, insulin-stimulated tyrosine phosphorylation of IRS-1 and p85 docking to IRS-1 was selectively abolished in skeletal muscle of MIRKO mice (Fig. 8.6b). Taken together, these data demonstrate that IR expression and early insulin-stimulated signaling is virtually abolished in muscle but remains normal in liver of MIRKO mice. To determine the consequence of reduced insulin signaling on downstream actions of insulin, we measured insulin-stimulated glucose transport in isolated skeletal muscle from WT, IR(*lox/lox*) and MIRKO mice. As shown in Figure 8.6c, in vitro insulin-stimulated glucose transport in soleus muscle from MIRKO mice was reduced by 75% as compared with controls. This reduction occurred without any change in basal glucose transport. These data indicate that abolishment of functional IR expression in skeletal muscle leads to a marked decrease in the ability of insulin to stimulate glucose transport.

To determine the physiological consequence of this severe, muscle-specific form of insulin resistance, we monitored triglycerides, cholesterol, and free fatty acids (FFAs) in serum, blood glucose concentrations and insulin levels in the fasted and fed state, and performed serial glucose and insulin tolerance testing. Over a range of ages from 4 to 11 months, MIRKO mice displayed marked hypertriglyceridemia that was 70% elevated ($p < 0.01$) and a 20% elevation ($p < 0.05$) in serum FFAs as compared with the control groups (Fig. 8.7a and 8.7b). These two findings are hallmarks of the previously described metabolic syndrome associated with insulin resistance, sometimes termed Syndrome X (Reaven, 1987).

It is surprising that despite the virtual lack of functional IR signaling in skeletal muscle, the MIRKO mice could maintain euglycemia for at least the first 11 months of life (Fig. 8.7c). Moreover, plasma insulin concentrations were indistinguishable between animals of each group (Fig. 8.7d). Despite the lack of insulin-stimulated signaling and the markedly reduced insulin-stimulated glucose transport in isolated muscles of MIRKO mice, these animals showed no signs of glucose intolerance or insulin resistance (Fig. 8.7e and 8.7f). Thus, the MIRKO mice utilize pathways distinct from those in skeletal muscle to maintain normoglycemia.

Even though body weights of MIRKO mice were not significantly different from controls, MIRKO mice were observed to have larger fat depots in several sites, including perirenal, subcutaneous, and perigonadal fat pads. One potential explanation for the increase in fat pad mass is that glucose is shunted from the muscle of MIRKO mice to adipose for metabolism. Reaven (1987) has hypothesized that elevated levels of circulating FFAs in humans and rodents with NIDDM result from insulin resistance at the level of serum FFA metabolism. Elevated FFA flux through the liver results in increased TG secretion rates, ultimately leading to hypertriglyceridemia. Our data indicate that the MIRKO mice, which have the sole genetic defect of insulin resistance in skeletal muscle, have a phenotype consistent with NIDDM-related serum lipoprotein abnormalities. These studies of MIRKO mice have led us to challenge the present model of NIDDM in which peripheral insulin resistance in muscle results in hypersecretion of insulin and other secondary phenomena resulting, eventually, in diabetes.

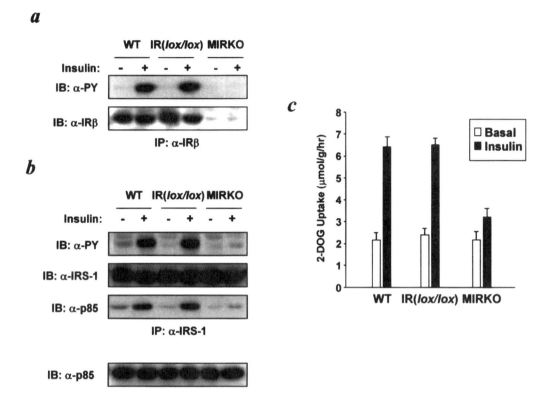

Figure 8.6. Insulin-stimulated events in MIRKO skeletal muscle are drastically diminished. (a) WT, IR(*lox/lox*) and MIRKO mice were anesthetized by intraperitoneal injection of pentobarbital and injected with either saline ($-$) or 5 IU of regular human insulin ($+$) via the inferior vena cava. Hind-limb skeletal muscle was removed after three minutes, and protein extracts were subjected to immunoprecipitation with an insulin receptor-specific antiserum followed by Western blot analysis with an anti-phosphotyrosine specific antibody (upper panel) or an insulin receptor-specific antiserum (lower panel). (b) The experiment was performed as described in (a) with the exception that immunoprecipitations were performed with an IRS-1-specific antiserum followed by Western blot analysis with an anti-phosphotyrosine specific antibody (upper panel), an IRS-1-specific antiserum (second panel from top) or an antiserum to the p85 regulatory subunit of PI 3-kinase (second panel from bottom). The bottom panel shows a Western blot analysis on the same extracts using an anti-p85 antiserum. (c) Fasted mice were sacrificed by cervical dislocation and 2-deoxyglucose transport was determined on isolated soleus muscle. Muscles from WT, IR(*lox/lox*), and MIRKO mice were untreated (*open bars*) or stimulated with 6.6 nM human regular insulin (*closed bars*).

7. THE ROLE OF THE INSULIN RECEPTOR IN THE ENDOCRINE PANCREAS

The regulation of insulin secretion from the β-cell is complex, and the molecular defects underlying the relative failure of the β-cell in type 2 diabetes are still unclear. One of the characteristic features of this defect is a failure of the β-cell to respond to a glucose stimulus while retaining its response to other secretagogues such as amino acids (Porte, Jr., 1991; Polonsky, 1995). Although insulin has been suggested to regulate its own secretion and has recently shown to regulate expression of some genes in the β-cell, the role of insulin action in the β-cell remains controversial (Elahi et al., 1982). One of the factors contributing to these conflicting data is the lack of biochemical evidence for the presence of a functional insulin receptor in the β-cells.

In the preceding section we have described the use of the Cre-loxP system (Bruning et al., 1998) to produce a muscle-specific knock-out of the insulin receptor. To directly assess the role of functional insulin receptors in the pancreatic β-cell in vivo, we have used a similar strategy.

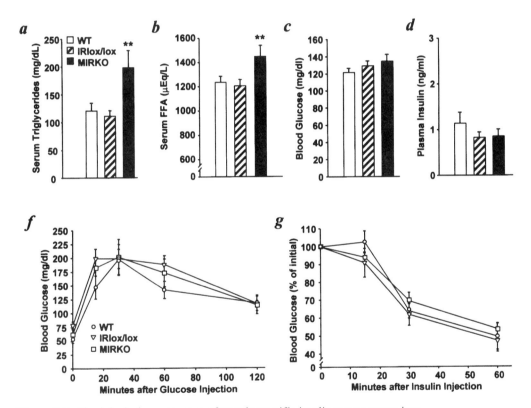

Figure 8.7. Physiological consequence of muscle-specific insulin receptor gene inactivation. Triglyceride (*panel a*) and free fatty acid (*panel b*) levels were determined on serum samples from overnight fasted, four-month-old WT (*open bar*), IR(*lox/lox*) (*hatched bar*), and MIRKO (*filled bar*) mice. Each bar represents the mean of at least 10 animals of each genotype +/− S.E.M. **, p < 0.01. Blood glucose (*panel c*) and serum insulin (*panel d*) concentrations were determined on venous blood samples from random-fed, four-month-old WT (*open bar*), IR(*lox/lox*) (*hatched bar*), and MIRKO (*filled bar*) mice. Each bar represents the mean of at least 10 animals of each genotype +/− S.E.M. Glucose tolerance (GTT, *panel e*) and insulin tolerance (ITT, *panel f*) tests were performed on four-month-month old WT (*circle*), IR(*lox/lox*) (*triangle*), and MIRKO (*square*) mice. Results of the GTT are expressed as mean blood glucose concentration +/− S.E.M. from at least eight animals of each genotype, whereas results of the ITT are expressed as mean percent of basal blood glucose concentration +/− S.E.M. from at least eight animals of each genotype.

Creation of the β-cell-Specific Insulin Receptor Knockout Mice

Mice carrying a Cre transgene driven by the β-cell specific rat insulin 2 promoter (Rip-Cre) were used to generate a β-cell specific knockout of the insulin receptor gene (Postic, 1999). These transgenic mice demonstrated specific Cre expression only in insulin producing β-cells, but not in non-β-cells or acinar cells as confirmed by double-label immunofluorescence (Fig. 8.8a). Furthermore, the ability of the Cre to induce pancreatic specific inactivation was demonstrated when these founder mice were bred with mice carrying a floxed

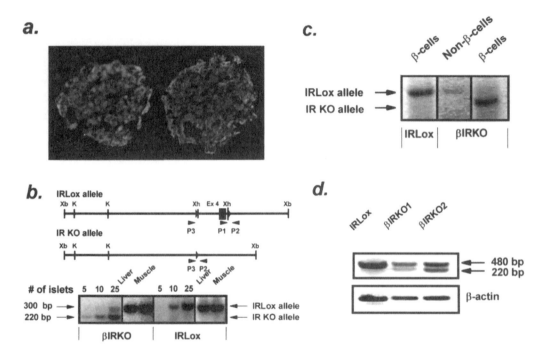

Figure 8.8. Assessment of Cre expression and insulin receptor recombination in βIRKO islets. (a) Immunofluorescent histochemical analysis of pancreas sections of a Cre-expressing mouse. Two individual islets are shown with Cre (red) staining localized to the insulin-producing β-cell (green). (b) Schematic representation of the IR(*lox/lox*) allele (IRLox, upper panel) showing the position of the different primers used in the PCR analysis. The knockout allele is shown below indicating the deletion of exon 4 in the event of recombination of the insulin receptor gene. The lower panel shows a representative PCR analysis of DNA prepared from 5, 10, and 25 islets, and from liver and skeletal muscle of βIRKO and IR(*lox/lox*) mice. The smaller 220 bp band is observed only in islets from βIRKO mice. (c) PCR analysis of DNA prepared from pure β and non-β cells obtained from flow cytometry using islets isolated from βIRKO and IR(*lox/lox*) mice. The smaller 220 bp band was observed in β cells from βIRKO islets (lane 3) but not in non-β cells from βIRKO islets (lane 2) or cells prepared from IR(*lox/lox*) islets (lane 1). (d) Reverse transcription-PCR analysis of RNA prepared from islets from one IR(*lox/lox*) and two different βIRKO mice. The RT reaction was carried out by using 1 μg of total RNA. A high-molecular-weight band (480 bp) is observed in all lanes, whereas a smaller band of 220 bp is observed only in islets from βIRKO mice, suggesting a recombination event. (See color plates.)

glucokinase gene (Postic, 1999). To create β-cell specific insulin receptor knockout (βIRKO) mice, the IR(*lox*/+) mice carrying the Cre transgene were bred with the IR(*lox*/+) mice. This breeding generated the βIRKO mice, for example, animals carrying the Cre transgene and homozygous for the IRlox allele and three littermate control groups—namely, homozygous IRlox mice [IR(*lox/lox*)]—to rule out the effect of introduction of *loxP* sites in the insulin receptor gene; mice carrying only the Rip-Cre transgene to rule out any effect of Cre expression on β-cell function; and WT controls. All animals were born normally at the expected Mendelian frequency, and no significant differences in body weights were observed between newborn (βIRKO mice and IR(*lox/lox*), Rip-Cre, and WT littermates up to 6 months. (βIRKO mice were fertile and produced normal-sized litters, and their development was not different from that observed in the control littermates (Kulkarni et al., 1999).

Assessment of the Effect of the Rip-Cre Transgene on the IRlox Allele

To evaluate the efficiency of recombination of the IRlox allele in the islet triggered by Cre expression, we used a PCR strategy using DNA prepared from islets isolated from βIRKO and IR(*lox/lox*) mice. A single 300-bp band was observed in islets from homozygous IRlox mice, indicating the presence of exon 4 (Fig. 8.8b). In contrast, with DNA prepared from islets from βIRKO mice, the major product was 220 bp in size and was the dominant band present in the lanes containing islet samples. A faint band was observed in addition to the 220-bp product in the sample lane with 25 islets in the βIRKO mice, indicating the presence of some un-recombined insulin receptor gene. This finding can be accounted for by the fact that β-cells make up approximately 80–85% of islet cells. Because the DNA prepared from the islets is a mixture of β-cell DNA, that from non-β islet cells (e.g., α-cells, δ-cells, and PP cells) as well as some integrated vascular tissues, one would expect at most ~80% of the islet DNA to exhibit a recombination of the insulin receptor gene. To confirm the specificity of the recombination event, we carried out IRLox PCR analysis of DNA prepared from pure β-cells as well as non-β cells obtained from dispersed islets using flow cytometry. A single 220-bp band was obtained in the lane loaded with β-cells from βIRKO islets, while a higher molecular weight band of 300 bp was evident only in the lanes with non-β cells from βIRKO islets and β-cells from IRLox islets (Fig. 8.8c). These data suggest a complete recombination of the insulin receptor gene specifically in the β-cells obtained from βIRKO islets. To confirm the recombination event in the islet β-cells, we also carried out RT-PCR using total RNA prepared from islets isolated from the IR(*lox/lox*) and βIRKO mice. A 480-bp PCR product was present in samples from both IR(*lox/lox*) and βIRKO mice, indicating the presence of normal insulin receptor mRNA. In addition, a smaller product of 220 bp was present only in the βIRKO sample confirming that recombination had occurred (Fig. 8.8d). Although these PCR reactions were not designed to be quantitative, the relative expression of intact receptor RNA in islet extracts of βIRKO mice as compared with the high level of recombination of DNA in these same islets suggests that most of the insulin receptor mRNA in the islet is derived from non- islet cells that have not undergone the recombination event. Thus, it seems likely that insulin receptors are expressed in both β- and non-β cells. The expression of the insulin receptor in other tissues of the mouse, including non-insulin responsive tissues like brain and insulin-responsive tissues such as the skeletal muscle, liver, and heart, was unaltered. Thus, the introduction of the *loxP* sites in the insulin receptor gene did not interfere with receptor expression, and knockout of the insulin receptor was specific to the β-cell in βIRKO mice.

8. ASSESSMENT OF THE EFFECT OF β-CELL SPECIFIC DISRUPTION OF THE INSULIN RECEPTOR

Effects on Insulin and Glucose Levels

To assess the effect of the β-cell insulin receptor knockout on glucose homeostasis, glucose and insulin levels were measured in the fasted and random-fed states. No significant differences were observed in blood glucose levels in the fasting and random-fed states in either male or female βIRKO mice as compared with those in the control groups up to six months of age. By 6 months, fasting insulin levels had also increased in all four groups of male mice, and a mild degree of hyperinsulinemia was observed in the βIRKO mice compared with the IR(*lox/lox*) and WT controls.

Decreased Acute Insulin Release to Glucose and Impaired Glucose Tolerance.
To evaluate the effect of β-cell-specific insulin receptor deletion on islet function, insulin release was measured in response to stimulation with the two major nutrient secretagogues of insulin—glucose and arginine. In both male and female control mice, a three- to four-fold increase in insulin secretion was observed 2 min after IP glucose injection (Fig. 8.9a and 8.9b). The insulin levels remained higher than baseline values for up to 30 min, indicating a second-phase response. In the βIRKO mice, the acute first-phase insulin secretory response to glucose was reduced by 85% in the females and was virtually absent in the males. Arginine stimulates acute insulin release by utilizing the transporter mCAT2 and by mechanisms independent of those used by glucose, although the final common pathways of secretion are the same (Weinhaus et al., 1997). Therefore, to determine the level of defect in glucose-stimulated insulin release, mice were also given an acute arginine challenge. In contrast to the response to glucose in both the βIRKO and control mice, a 5- to 6-fold potentiation of the glucose-stimulated acute insulin was released in response to arginine (Fig. 8.9c and 8.9d). Thus the βIRKO mice show a near- complete loss of acute first-phase insulin release specific to glucose while maintaining the response to arginine, suggesting that a functional insulin receptor and/or insulin response is requisite for a normal glucose-stimulated secretory response by the β-cell. This finding is unlikely to be a non-specific consequence of the genetic manipulation because the control IR(*lox/lox*) and Cre mice showed normal secretory responses. To assess the impact of altered first-phase insulin release and the ability of the mice to dispose of a glucose load, glucose disposal was assessed by glucose tolerance tests (Fig. 8.10). As noted above, at all ages up to 6 months,

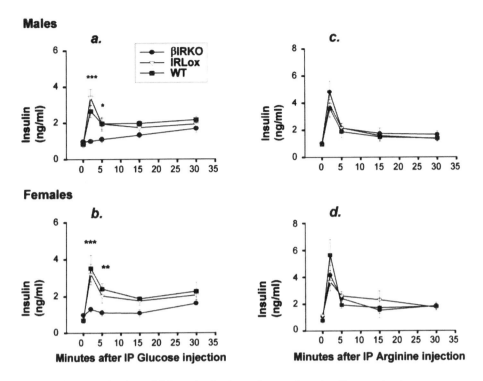

Figure 8.9. βIRKO mice exhibit a selective loss of acute phase insulin secretion in response to glucose but not to arginine. Glucose-stimulated acute phase insulin secretion was tested by intra-peritonal injection of glucose (3 mg/g body weight) in three- to four-month-old male (a) and female (b) WT (*filled squares*), IR(*lox/lox*) (*open triangles*), and βIRKO (*filled circles*) mice. Blood samples were collected at the indicated time intervals from tail-vein samples for insulin RIA. An absent or significantly blunted response was observed in male and female βIRKO mice, respectively, compared with controls. Values are means SEM, ($n = 3$ to 14). *** $p < 0.001$, ** $p < 0.01$, * $p < 0.05$; βIRKO versus WT or IR(*lox/lox*). Arginine-stimulated acute-phase insulin secretion was tested by IP injection of L-arginine (0.3 mg/g body weight) as described under methods. No significant differences were observed between the groups in male (c) or female (d) mice. Values are means SEM, ($n = 7$ to 8).

fasting glucose levels were the same in the βIRKO mice as in the controls. Upon glucose challenge, however, both male and female βIRKO mice showed significantly higher glucose levels at 15 and 30 min after the injection compared with controls at 2 months of age (Fig. 8.10a and 8.10b). Glucose tolerance continued to worsen in the βIRKO mice with age, and by six months, glucose levels were more than twice as high in the βIRKO mice as in controls in both males and females (Fig. 8.10e and 8.10f).

Morphology and Insulin Content of The Pancreas. To evaluate if the alterations in endocrine pancreatic function are associated with any changes in morphology of the islets, we carried out immunohistochemical studies on the pancreas. Analysis of sections of pancreas using a cocktail of antibodies on

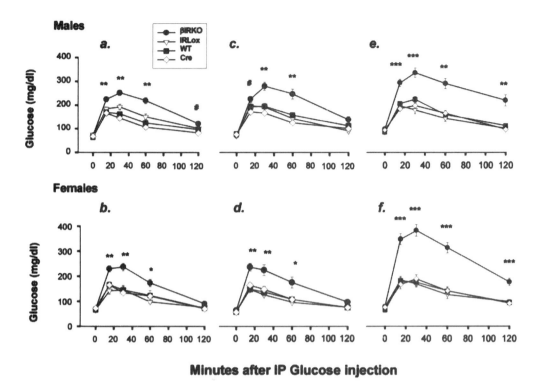

Figure 8.10. βIRKO mice demonstrate a progressively impaired glucose tolerance. The ability to handle a glucose load was assessed by carrying out a glucose tolerance test at two, four, and six months of age in WT (*filled squares*), IR(*lox/lox*) (*open triangles*), Cre (*open diamonds*), and βIRKO (*filled circles*) male (*a, c, e*) and female (*b, d, f*) mice. Blood glucose was determined by tail-vein sampling immediately before (0 min) and 15, 30, 60, and 120 mins after the IP injection of glucose (2 g/kg body weight). An age-dependent glucose intolerance was observed in both male and female mice. Results are expressed as mean blood glucose concentrations $\pm$ SEM, ($n = 10$ to 25). *** $p < 0.001$, ** $p < 0.01$, * $p < 0.05$ βIRKO *vs* WT or IR(*lox/lox*) or Cre. # $p < 0.05$ βIRKO versus WT or Cre.

non-β-cell hormones showed no apparent differences in islet size or in the ratio of β cells to non-β cells at 2 months of age (Fig. 8.11a, left panel). Assessment of insulin content in the pancreas showed comparable levels at 2 months of age consistent with the islet morphology (Fig. 8.11b, left panel). However, 4-month-old βIRKO mice showed a modest decrease in islet size when compared with controls (Fig. 8.11a, right panel). Insulin content at this age showed a small increase in control mice compared with 2-month levels, whereas the insulin content remained unchanged in the βIRKO mice. Results also showed a lower insulin content in 4-month-old βIRKO mice as compared with controls (Fig. 8.11b, right panel). Furthermore, examination of other factors known to influence the acute-phase insulin release such as glut2 was relatively normal in the islets and there was no difference in β-cell morphology at the electron microscope level. It appears likely that the insulin receptor is essential for normal glucose sensing or in the specific secretory machinery involved in glucose-stimulated insulin release.

Taken together, these data demonstrate that a mouse with a specific deletion of the β-cell insulin receptor exhibits a selective loss of the acute-phase secretion in response to glucose and progressively impaired glucose tolerance over 5 months. These findings provide direct evidence of a functional role for the insulin receptor in the islet β-cell in the maintenance of glucose homeostasis in vivo and suggest that insulin resistance at the level of the β-cell may be a significant factor in the development of a loss of glucose-stimulated insulin secretion by the β-cells. Thus, a novel unifying hypothesis can be formulated for type 2 diabetes in which insulin resistance at the β-cell level coupled with insulin resistance at the periphery could result in the classical pathophysiological findings of this disease. Whereas it is unlikely that in humans this condition is due to genetic alterations in the insulin receptor, acquired alterations in the receptor or genetic and acquired alterations in the post-receptor signaling steps could provide a mechanism for this complex disease syndrome.

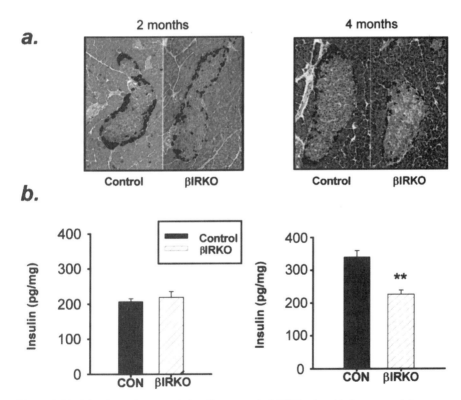

Figure 8.11. Islet size and pancreatic insulin content in βIRKO mice. (a). Immunostaining for non-β-cell hormones was carried out in pancreatic sections from two- and four-month-old control [WT and IR(*lox/lox*)] and βIRKO mice by using a cocktail antibody to glucagon, somatostatin, and pancreatic polypeptide. Representative sections are shown from a control (CON) and βIRKO mouse. (b) Pancreatic insulin content was measured in acid-ethanol extracts from control [WT and IR(*lox/lox*)] and βIRKO mice using a RIA kit. Values are means SEM, (*n* = 3 to 5).** *p* < 0.01 βIRKO versus Control. (See color plates.)

8. CONCLUSION AND FUTURE DIRECTIONS

Although type 2, or adult-onset, diabetes mellitus has been recognized as a disease for more than 2000 years, the primary defect that accounts for this disorder has not been determined. Although epidemiological studies implicate a polygenic inheritance and genetic heterogeneity—as well as strong influences by environmental factors such as diet, obesity, and level of activity (Warram et al., 1995) —longitudinal studies have shown that the first detectable defect in the pathogenesis of NIDDM is resistance to insulin-stimulated glucose uptake in skeletal muscle (Martin et al., 1992; Lillioja et al., 1993; Kahn, 1994). In the present chapter we have described some of the state-of-the-art techniques that have been used to specifically address the role of the skeletal muscle in the pathogenesis of type 2 diabetes. From the results discussed above, the use of conditional gene targeting has provided surprising yet important information on the contribution of the skeletal muscle to several features of the metabolic syndrome associated with type 2 diabetes but not glucose tolerance. Furthermore, the unexpected significance of the insulin receptor in the pancreatic β-cells as an important player in glucose homeostasis indicates the complexity of the disease process. Both these models provide strong evidence for the need to reassess the pathogenesis underlying type 2 diabetes. Whereas conditional gene targeting of other tissues, including the liver and adipose, will no doubt prove significant in further deciphering the pathogenesis of type 2 diabetes, the strategy of tissue-specific knockouts in a temporal manner with time-dependent promoters may prove more informative in the study of disease processes with an adult onset such as type 2 diabetes.

REFERENCES

Accili D, Drago J, Lee EJ, Johnson MD, Cool MH, Salvatore P, et al. (1996): Early neonatal death in mice homozygous for a null allele of the insulin receptor gene. Nat Genet 12:106–109.

Almind K, Bjorbaek C, Vestergaard H, Hansen T, Echwald SM, Pedersen O (1993): Amino acid polymorphisms of insulin receptor substrate-1 in non-insulin-dependent diabetes mellitus. Lancet 342:828–832.

Andrews BJ, Proteau GA, Beatty LG, Sadowski PD (1985): The FLP recombinase of the 2 micron circle DNA of yeast: interaction with its target sequences. Cell 40:795–803.

Araki E, Lipes MA, Patti ME, Brüning JC, Haag BL III, Johnson RS (1994): Alternative pathway of insulin signalling in mice with targeted disruption of the IRS-1 gene. Nature 372:186–190.

Araki K, Araki M, Miyazaki J, Vassalli P (1995): Site-specific recombination of a transgene in fertilized eggs by transient expresson of cre recombinase. PNAS USA 92:160–164.

Bruning JC, Michael MD, Winnay JN, Hayashi T, Horsch D, Accili D, et al. (1998): A muscle-specific insulin receptor knockout challenges the current concepts of glucose disposal and NIDDM pathogenesis. Mol Cell 2:559–569.

Bruning JC, Winnay J, Bonner-Weir S, Taylor SI, Accili D, Kahn CR (1997): Development of a novel polygenic model of NIDDM in mice heterozygous for IR and IRS-1 null alleles. Cell 88:561–572.

Chang PY, Benecke H, Le Marchand-Brustel Y, Lawitts JA, Moller DE (1994): Expression of a dominant-negative mutant human insulin receptor in the muscle of transgenic mice. J Biol Chem 269:16034–16040.

Cheatham B, Kahn CR (1992): Cysteine 647 in the insulin receptor is required for normal covalent interaction between alpha- and beta-subunits and signal transduction. J Biol Chem 267:7108–7115.

Cheatham B. Kahn CR (1995): Insulin action and the insulin signaling network. Endocr Rev 16:117–142.

Cheatham B, Shoelson SE, Yamada K, Goncalves E, Kahn CR (1993): Substitution of the erbB-2 oncoprotein transmembrane domain activates the insulin receptor and modulates the action of insulin and insulin-receptor substrate 1. PNAS USA 90:7336–7340.

Cheatham B, Vlahos CJ, Cheatham L, Wang L, Blenis J, Kahn CR (1994): Phosphatidylinositol 3-kinase activation is required for insulin stimulation of pp70 S6 kinase, DNA synthesis, and glucose transporter translocation. Mol Cell Biol 14:4902–4911.

Chen J, Kubalak SW, Chien KR (1998): Ventricular muscle-restricted targeting of the RXRalpha gene reveals a non-cell-autonomous requirement in cardiac chamber morphogenesis. Development 125:1943–1949.

Czech MP (1981): Insulin Action. Am J Med 70 142–150.

Dobie KW, Lee M, Fantes JA, Graham E, Clark AJ, Springbett A (1996): Variegated transgene expression in mouse mammary gland is determined by the transgene integration locus. PNAS USA 93, 6659–6664.

Dooley TP, Miranda M, Jones NC, DePamphils ML (1989): Transactivation of the adenovirus EIIa promoter in the absence of adenovirus EIA protein is restricted to mouse oocytes and preimplantation embryos. Development 107 945–956.

Dymecki SM (1996): FLP recombinase promotes site-specific DNA recombination in embryonic stem cells and transgenic mice. PNAS USA 93, 6191–6196.

Elahi D, Nagulesparan M, Hershcopf RJ, Muller DC, Tobin JD, Blix PM, Rubenstein AH, Unger RH, Andres R (1982): Feedback inhibition of insulin secretion by insulin: Relation to the hyperinsulinemia of obesity. N Engl J Med 306 1196–1202.

Fiering S, Epner E, Robinson K, Zhuang Y, Telling A, Hu M, Martin DI, Enver T, Ley TJ, Groudine M (1995): Targeted deletion of 5′HS2 of the murine beta-globin LCR reveals that it is not essential for proper regulation of the beta-globin locus. Genes Dev 15:2203–2213.

Flier JS, Kahn CR, Roth J, Bar RS (1975): Antibodies that impair insulin receptor binding in an unusual diabetic syndrome with severe insulin resistance. Science 190:63–65.

Gagneten S, Le Y, Miller J, Sauer B (1997): Brief expression of a GFP cre fusion gene in embryonic stem cells allows rapid retrieval of site-specific genomic deletions. Nucleic Acids Res 25:3326–3331.

Gammeltoft S, Fehlmann M, Van Obberghen E (1985): Insulin receptors in the mammalian central nervous system: binding characteristics and subunit structure. Biochimie 67:1147–1153.

Goncalves E, Yamada K, Thatte HS, Backer JM, Golan DE, Kahn CR, Shoelson SE (1993): Optimizing transmembrane domain helicity accelerates insulin receptor internalization and lateral mobility. PNAS USA 90:5762–5766.

Gu H, Marth JD, Orban PC, Mossmann H, Rajewsky K (1994): Deletion of a DNA polymerase beta gene segment in T cells using cell type-specific gene targeting. Science 265, 103–106.

Hanks SK, Quin AM, Hunter T (1988): The protein kinase family: conserved features and deduced phylogeny of the catalytic domain. Science 241, 42–52.

Hoess R, Abremski K, Irwin S, Kendall M, Mack A (1990): DNA specificity of the cre recombinase resides in the 25 kDa carboxyl domain of the protein. J Mol Biol 216, 873–882.

Imai Y, Fusco A, Suzuki Y, Lesniak MA, D'Alfonso R, Sesti G, Bertoli A, Lauro R, Accili D, Taylor SI (1994): Variant sequences of insulin receptor substrate-1 in patients with noninsulin dependent diabetes mellitus. J Clin Endocrinol Metab 79:1655–1658.

Johnson JE, Wold BJ, Hauschka SD (1989): Muscle creatine kinase sequence elements regulating skeletal and cardiac muscle expression in transgenic mice. Mol Cell Biol 9 3393–3399.

Joshi RL, Lamothe B, Cordonnier N, Mesbah K, Monthioux E, Jami J, Bucchini C (1996): Targeted disruption of the insulin receptor gene in the mouse results in neonatal lethality. EMBO J. 15:1542–1547.

Kahn CR (1994): Insulin action, diabetogenes, and the cause of type II diabetes (Banting Lecture). Diabetes 43:1066–1084.

Kahn CR. White MF (1988): The insulin receptor and the molecular mechanism of insulin action. J Clin Invest 82:1151–1156.

Katz EB, Stenbit AE, Hatton K, DePinho RA, Charron MJ (1995): Cardiac and adipose tissue abnormalities but not diabetes in mice deficient in GLUT4. Nature 377:151–155.

Kozak M (1999): Influences of mRNA secondary structure on initiation by eukaryotic ribosomes. PNAS USA 83:2850–2854.

Krempler F, Hell E, Winkler C, Breban D, Patsch W (1998): Plasma leptin levels: interaction of obesity with a common variant of insulin receptor substrate-1. Arterioscler Thromb Vasc Biol 181686–1690.

Kulkarni RN, Brüning JC, Winnay JN, Postic C, Magnuson MA, Kahn CR (1999): Tissue-specific knockout of the insulin receptor in pancreatic β cells creates an insulin secretory defect similar to that in type 2 diabetes. Cell 96:329–339.

Kulkarni RN, Winnay JN, Daniels M, Brüning JC, Filer SN, Hanahan D, Kahn CR (1999): Altered function of insulin receptor substrate-1-deficient mouse islets and cultured β-cell lines. J Clin Invest 104:R69–R75.

Laakso M, Malkki M, Kekalainen P, Kuusisto J, Deeb SS (1994): Insulin receptor substrate-1 variants in non-insulin-dependent diabetes. J Clin Invest 94:1141–1146.

Lakso M, Pichel JG, Gorman JR, Sauer B, Okamoto Y, Lee E, Alt FW, Westphal H (1996): Efficient in vivo manipulation of mouse genomic sequences at the zygote stage. PNAS USA 93:5860–5865.

Lakso M, Sauer B, Mosinger B, Jr, Lee EJ, Manning RW, Yu SH, Mulder KL, Westphal H (1992): Targeted oncogene activation by site-specific recombination in transgenic mice. PNAS USA 896232–6236.

Lauro D, Kido Y, Castle AL, Zarnowski MJ, Hayashi H, Ebina Y, Accili D (1998): Impaired glucose tolerance in mice with a targeted impairment of insulin action in muscle and adipose tissue [see comments]. Nat Genet 20294–298.

Lillioja S, Mott DM, Spraul M, Ferraro R, Foley JE, Ravussin E, Knowler WC, Bennett PH, Bogardus C (1993): Insulin resistance and insulin secretory dysfunction as precursors of non-insulin-dependent diabetes mellitus: prospective studies of Pima Indians. N Engl J Med 329:1988–1992.

Lobe CG, Koop KE, Kreppner W, Lomeli H, Gertsenstein M, Nagy A (1999): Z/AP, a double reporter for cre-mediated recombination. Dev Biol. 208:281–292.

Mao X, Fujiwara Y, Orkin SH (1999): Improved reporter strain for monitoring cre recombinase-mediated DNA excisions in mice. PNAS USA 96:5037–5042.

Martin BC, Warram JH, Krolewski AS, Bergman RN, Soeldner JS, Kahn CR (1992): Role of glucose and insulin resistance in development of Type II diabetes mellitus: results of a 25-year follow-up study. Lancet 340:925–929.

Moller DE, Chang PY, Yaspelkis BB, III, Flier JS, Wallberg-Henriksson H, Ivy JL (1996): Transgenic mice with muscle-specific insulin resistance develop increased adiposity, impaired glucose tolerance, and dyslipidemia. Endocrinology 137: 2397–2405.

Papaioannou V, Johnson R. Production of chimeras and genetically defined offspring from targeted ES cells. In Gene Targeting: A Practical Approach; AL Joyner, Ed. Oxford University Press: Oxford, England, pp 107–146.

Polonsky KS (1995): Lilly Lecture 1994. The beta-cell in diabetes: from molecular genetics to clinical research. Diabetes 44, 705–717.

Porte D, Jr. (1991): Banting lecture 1990. β-cells in type II diabetes mellitus. Diabetes 40:166–180.

Postic C (1999): Cell-specific roles of glucokinase in glucose homeostasis as determined by liver and pancreatc β cell-specific gene knock-outs using Cre recombinase. J Biol Chem 274:305–315.

Reaven GM (1987): Non-insulin-depndent diabetes mellitus, abnormal lipoprotein metabolism and atherosclerosis. Metabolism 36:1–8.

Sauer B, Henderson N (1990): Targeted insertion of exogenous DNA into the eukaryotic genome by the cre recombinase. New Biol. 2:441–449.

Schlessinger J (1988): Signal transduction by allosteric receptor oligomerization. Trends Biochem Sci 13:443–447.

Schwenk F, Baron U, Rajewsky K (1995): A cre-transgenic mouse strain for the ubiquitous deletion of loxP-flanked gene segments including deletion in germ cells. Nucleic Acid Res 23:5080–5081.

Seino S, Bell GI (1989): Alternative splicing of human insulin receptor messenger RNA. Biochem Biophys Res Comm 159:312–316.

Sternberg N, Sauer B, Hoess R, Abremski K (1986): Bacteriophage P1 cre gene and its regulatory region. Evidence for multiple promoters and for regulation by DNA methylation. J Mol Biol 187:197–212.

Sunaga S, Maki K, Komagata Y, Ikuta K, Miyazaki JI (1997): Efficient removal of loxP-flanked DNA sequences in a gene-targeted locus by transient expression of cre recombinase in fertilized eggs. Mol Reprod Dev 46:109–113.

Tamemoto H, Kadowaki T, Tobe K, Yagi T, Sakura H, Hayakawa T, et al. (1994): Insulin resistance and growth retardation in mice lacking insulin receptor substrate-1. Nature 372:182–186.

Taylor SI (1992): Lilly Lecture 1992. Molecular mechanisms of insulin resistance-Lessons from patients with mutations in the insulin receptor gene. Diabetes 41:1473–1490.

Terauchi Y, Tsuji T, Satoh S, Minoura H, Murakami K, Okuno A et al. (1999): Increased insulin sensitivity and hypoglycaemia in mice lacking the p85 alpha subunit of phosphoinositide 3-kinase. Nat Genet 21:230–235.

Trask RV, Billadello JJ (1990): Tissue-specific distribution and developmental regulation of M and B creatine kinase mRNAs. Biochim Biophys Acta 1049:182–188.

Ullrich A, Bell JR, Chen EY, Herrera R, Petruzzelli LM, Dull TJ, et al. (1985): Human insulin receptor and its relationship to the tyrosine kinase family of oncogenes. Nature 313:756–761.

Warram JH, Rich SS, Krolewski AS. Epidemiology and genetics of diabetes mellitus. In Joslin's Diabetes Mellitus; CR Kahn, GC Weir, Eds. Lea & Febiger: Philadelphia.

Weinhaus AJ, Poronnik P, Tuch BE, Cook DI (1997): Mechanisms of arginine-induced increase in cytosolic calcium concentration in the beta-cell line NIT-1. Diabetologia 40:374–382.

Withers DJ, Gutierrez JS, Towery H, Burks DJ, Ren JM, Previs S, et al. (1998): Disruption of IRS-2 causes type 2 diabetes in mice. Nature 391:900–903.

CHAPTER 9

TARGETED DOMINANT NEGATIVE MUTATIONS

WILLIAM J. LAROCHELLE, GIULIA CELLI,
and GLENN MERLINO*

I. INTRODUCTION: THE DOMINANT NEGATIVE APPROACH TO IN VIVO RECEPTOR FUNCTION

Investigations using genetically engineered mice are progressively revealing the functional role of peptide growth factors and their receptors in normal growth and development, as well as in disease. Gain-of-function transgenic mice inappropriately expressing specific ligands and/or receptors now serve as highly useful animal models for a number of human diseases, including breast carcinoma and malignant melanoma (see reviews by Merlino, 1994; Amundadottir et al., 1996; Chin et al., 1998, any others in this book). More recently, the advent of efficient methods to induce mammalian homologous recombination and the development of totipotent murine embryonic stem cell lines have made possible the generation of mice carrying inactive or crippled genes (see reviews in this book). Although frequently providing remarkable insights into gene function, this loss-of-function approach has notable deficiencies in certain situations. For example, it is not uncommon to observe the association of early embryonic lethality with null mutations in critical genes, dramatically limiting the use of such mouse models. More frustrating still, perhaps, is the failure to induce an overt phenotype in so-called "knock-out animals" in which one member of a family of functionally redundant genes, such as those encoding the fibroblast growth factors (FGFs) and their receptors (FGFRs), has been inactivated.

Many of these inadequacies can be overcome by using transgenic mice expressing dominant negative variants in a targeted, regulative fashion. A dominant negative mutation can be defined genetically as one leading to the expression of a mutant gene product capable of inhibiting the function of the corresponding wild-type gene product in heterozygotes. In the case of receptors, dominant negative variants can be designed that are capable of disrupting multiple family

*Corresponding author.

Genetic Manipulation of Receptor Expression and Function,
Edited by Domenico Accili.
ISBN 0-471-35057-5   Copyright © 2000 Wiley-Liss, Inc.

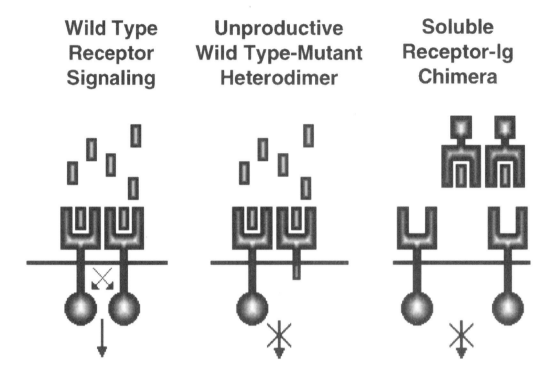

Wild Type Receptor Signaling   **Unproductive Wild Type-Mutant Heterodimer**   **Soluble Receptor-Ig Chimera**

Figure 9.1. Schematic representation of the postulated mechanisms by which dominant negative receptors function. Normally (left panel), ligand (rectangles) binding induces dimerization of native RTKs, autophosphorylation, and activation of downstream signal transduction pathways. In the case of the prototypical dominant negative receptor (center panel), the truncated kinase-deficient mutant remains tethered to the membrane and in-activates native receptors by forming unproductive heterodimers. In contrast, soluble re-ceptor-Ig chimeras (right panel) are secreted as dimers, accumulate, and may have a competitive edge by binding ligand before activation of native cell-surface receptors.

members, mitigating the problems associated with functional redundancy. Be-sides overcoming limitations associated with gene-targeting studies, dominant negative receptors offer great potential as versatile and efficacious therapeutic agents (see below). For receptor tyrosine kinases (RTKs), the prototype dominant negative receptor retains the extracellular binding domain and the hydrophobic membrane spanning domain but is usually truncated rendering it kinase-deficient (Fig. 9.1). Dominant negative suppression of RTKs, such as the epidermal growth factor receptor (EGFR) and FGFR, whose activation is associated with dimeriza-tion, is thought to be achieved through unproductive heterodimer formation by defective and wild-type receptor pairing (Kashles et al., 1991).

In a landmark paper, Amaya et al. (1991) successfully engineered and ex-pressed in *Xenopus* embryos a truncated, membrane-bound FGF receptor mu-tant, disrupting mesoderm formation and demonstrating an essential role in early embryogenesis. Since then, an impressive array of studies has used simi-lar dominant negative receptor variants, both naturally occurring and artifi-cially created, to examine RTK function in vivo. Truncated FGFRs have now successfully been targeted with disruptive effect to the epidermis, lung-bud epithelium, lens, retina, cerebellar neurons, mammary gland, and pre-implantation embryos (Werner et al., 1993, 1994 Peters et al., 1994; Robinson et al., 1995; Chow et al., 1995; Campochiaro et al., 1996; Saffell et al., 1997; Jackson et al., 1997; Chai et al., 1998). Other RTKs that have demonstrated dominant negative activity as truncated isoforms in vivo include the EGFR and the vascular endothelial growth factor (VEGF) receptor (Flk-1) (Murillas et al., 1995; Xie et al., 1997; Millauer et al., 1994). Alternatively, dominant negative variants have been used in vivo, in which a kinase inactivating point mutation inhibits RTK activity. For example, transgenic expression of the kinase-defective c-*kit* mutant from the W42 allele at the white-spotting locus (*W*) recapitulated some of the *W* phenotypes, and it confirmed that this mutation operates in a dominant negative fashion (Ray et al., 1991). Tek and the insulin receptor signaling pathways have also been disrupted in transgenic mice by using RTK missense mutants (Dumont et al., 1994; Chang et al., 1995).

Besides RTKs, other types of receptors have also been the subject of dominant negative analysis in vivo, including serine/threonine kinase and retinoic acid receptors. Transgenic mice ectopically expressing a kinase-deficient transforming growth factor beta (TGFβ) type II receptor have demonstrated roles for TGFβ signaling in growth and differentiation of pancreatic acinar cells, keratinocytes, chondrocytes, osteocytes, mammary gland epithelium, as well as in carcinogenesis of the skin, lung, and mammary gland (see review by Letterio and Bottinger, 1998; Bottinger et al., 1997). Dominant negative retinoic acid receptor mutants have uncovered developmental roles for retinoid signaling in the skeleton and skin (Damm et al., 1993; Imakado et al., 1995; Saitou et al., 1995; Feng et al., 1997; Yamaguchi et al., 1998).

Despite the reported success of these transgenic studies, the major challenge to the dominant negative receptor approach in animal experiments and in human therapeutics is overcoming the large stoichiometric excess of the mutant form required to effectively compete for ligand-binding with the wild-type receptor, and thereby disrupt normal signaling. For the typical kinase-deficient, membrane tethered receptor this required excess is thought to be ≥10-fold (Ueno et al., 1992). Alternatively, more stable dominant negative agents designed to have an advantage in binding ligand in vivo, relative to their wild-type counterparts, would be highly desirable. In this regard, soluble dominant negative receptor chimeras currently offer great promise. In this review we will discuss how dominant negative receptor-Ig chimeras are characterized in vitro, how they can be used to address basic questions about receptor biology and function in vivo, and how their potential can be exploited in the diagnostic and therapeutic arenas.

Rationale for Dominant Negative Receptor-IgG Fc Domain Chimeras

Receptor-immunoglobulin (Ig) fusion proteins, or immunoadhesins, are highly effective as dominant negative receptors. Our laboratory first used receptor-IgG chimeras as ligand probes, as well as to map ligand-binding determinants of Ig-like receptors (Cheon et al., 1994; Heidaran et al., 1995; LaRochelle et al., 1995). In undertaking these studies, we reasoned that expression of individual receptor Ig-like domains within the context of an IgG molecule, whose Fc portion and flexible hinge regions were preserved, might provide structural

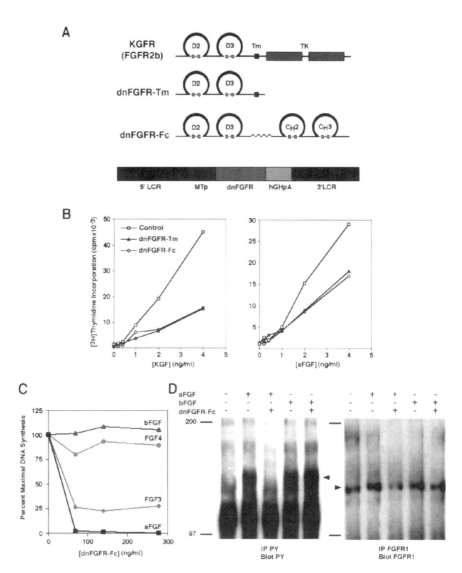

Figure 9.2. Structure and in vitro activity of dominant negative KGFR/FGFR2b mutants. (**A**) Schematic diagram of native and dominant negative receptors, as well as transgene construct. (D2, D3), Ig-like loops 2 and 3 of KGFR extracellular ligand-binding domain; (Tm), transmembrane region; (TK), split tyrosine kinase domain; (C_H2, C_H3), constant regions 2 and 3 of Ig heavy chain, and the hinge region (zigzag line). The transgene consists of the mouse metallothionein (MT) promoter and its flanking locus control regions (LCR) (see Palmiter et al., 1993), the human growth hormone polyadenylation signal (hGHpA), and the cDNA of either the truncated or chimeric dnFGFR. (**B**) Balb/MK cells were transfected with either dnFGFR-Tm or dnFGFR-Fc, and the activity of each mutant measured as inhibition of the mitogenic response to either aFGF and KGF. (**C**) The ligand-binding specificity of the soluble chimera was determined by the ability of increasing amounts of purified dnFGFR-Fc to block the proliferative response of NIH 3T3 cells to distinct FGFs. (**D**) Starved NIH 3T3 cells were treated with aFGF or bFGF in the presence or absence of purified dnFGFR-Fc. Cellular extracts were immunoprecipitated (IP) and subsequently probed with anti-PY or anti-FGFR1. Modified with permission from Celli et al., 1998.

constraints necessary for maintenance of the receptor's ligand-binding function. This dominant negative approach has proven useful for mapping the ligand-binding domains and specifically neutralizing the biological activity of the β platelet-derived growth-factor receptor (PDGFR) as well as the keratinocyte growth-factor receptor (KGFR or FGFR2b, Fig. 9.2A), and is now being exploited in vivo to gain insight into the biologic and pathologic roles of RTKs.

The Ig superfamily encompasses molecules involved in functions as diverse as immunorecognition, cell adhesion, antigen presentation, and MHC recognition (Williams and Barclay, 1988). Individual domains of the IgG molecule appear to encode distinct functions as well. For example, the first Ig loop of the heavy and light chains contain the determinants or idiotypes responsible for antigen/antibody interaction (Davies et al., 1988). The second IgG loop, CH1, functions in complement C3 binding (Shohet et al., 1993) whereas C_H2, the third IgG loop, tightly interacts with C1q as well as Fc receptors (Duncan et al., 1988; Lund et al, 1992) and mediates the plasma clearance of IgG (Burton, 1985). Finally, C_H3 appears important for dimerization and stability of the IgG molecule (Burton, 1985). Therefore, by analogy with the IgG molecule, different Ig-like domains of a prototypical Ig-like receptor may have evolved to exhibit very distinct functions.

Unique biologic events result as growth factors discriminate among specific members of the Ig-like receptor superfamily whose signaling plays essential roles in intercellular communication. Among the many Ig-like receptor superfamily members are the PDGF and FGF receptor multigene families (Yarden et al., 1986; Givol and Yanyon, 1992). We first hypothesized that individual receptor Ig-like domains might, as in the Ig molecule, specify unique functions. Therefore, expression of individual receptor Ig-like domains within the context of the IgG Fc and flexible hinge domains would favor high-affinity ligand-binding activity. The biologic stability of the chimera might also be enhanced. In addition to defining ligand-receptor interactions, our approach also has general applications for in vivo antagonists as well as searching for specific agonists and antagonists of receptor function. Such chimeric molecules would also have the advantage of being secreted from transfected cells, as well as detection and purification from medium utilizing conventional methods (Ey et al., 1978).

The functional in vitro utility of chimeric receptor proteins fused to immunoglobulins has been clearly demonstrated in many recent reports. The use of bifunctional molecules engineered as IgG heavy-chain chimeras was first reported for the CD4 immunoadhesin (Capon et al., 1989). Since then, receptor-Ig chimeras have been used for structural and functional studies of the Ig-like superfamily of receptors (Williams and Barclay, 1988), including the T-cell antigen receptor (Eilat et al., 1992; Gascoigne, 1987), VEGFR/KDR (Kaplan et al., 1997), Flt3/Flk-2 (Lyman et al., 1993), c-Kit (Liu et al., 1993), KGFR (Cheon et al., 1994), and β PDGFR (Heidaran et al., 1995). The functional properties of other receptors such as the Lymphotoxin β receptor (Ettinger et al., 1996), tumor necrosis factor receptor (Howard et al., 1993), TGFβ receptor (Isaka et al., 1999), and hepatocyte growth-factor receptor (Mark et al., 1992) have been analyzed by means of Ig fusion proteins as well. The three receptor subtypes for the natriuretic peptides, when fused to the Fc portion of human IgG gamma chain, were quantitatively and qualitatively indistinguishable from the native receptor (Bennett et al., 1991).

Biochemical Characterization of Receptor-Ig Chimeras

Before one considers the use of a specific receptor-Ig chimera in vivo, a thorough understanding of its mechanism of action, and proof that it retains the expected biologic properties of the native receptor, is essential to ensure a successful outcome. For example, our immunoadhesin technology proved generally applicable to map β PDGFR ligand-binding domains as well as to identify receptor antagonists (Heidaran et al., 1995). Previous studies using α and β PDGFR chimeras had shown that PDGF AA binding to the α PDGFR mapped to domains 1 to 3 (D1-3). However, because both PDGFRs bound PDGF BB, structural domains of the β PDGFR involved in PDGF BB binding could not be resolved (Heidaran et al., 1990; Yu et al., 1994). As an alternative approach, we constructed fusion proteins comprising β PDGFR Ig-like domains 1 to 3 (D1-3) and the IgG HFc domain. Scatchard analysis revealed that PDGF BB possessed high-affinity receptor-like binding to β PDGFR(D1-3)-Fc. No other individual Ig-like domains bound PDGF BB with significant affinity. Thus, β PDGFR Ig-like domains 1 to 3 were concluded to be sufficient for high-affinity PDGF BB binding (Heidaran et al., 1995). Of note, the VEGF binding domain of KDR has also been localized to Ig-like domains 1–3 by using Ig Fc chimeras (Kaplan et al., 1997).

Using the receptor-Ig chimeric approach, we were also successful in developing a rapid and sensitive β PDGFR-Fc immunosorbent assay for receptor antagonist screening. In the ELISA-like format, refolded human α PDGFR domains 1–3 inhibited β PDGFR-Fc binding to PDGF BB, suggesting that α PDGFR D1–3 were sufficient for high-affinity PDGF BB binding. Further studies with PDGFR receptor chimeras confirmed these results and suggested that D2 and 3 were required for high-affinity α PDGFR binding to PDGF AA (Mahadevan et al., 1995).

Mapping of ligand-binding determinants to Ig-like receptors led to a comparative study of KGF, also known as FGF7, and aFGF binding to the KGFR/FGFR2b as well (Miki et al, 1991). KGF binds and activates the KGFR (reviewed in Rubin et al., 1995), a membrane-spanning tyrosine kinase generated by alternative splicing of FGFR2 (one of four members of the FGFR gene family). The FGFR extracellular domain consists of three Ig-like domains (designated D1, D2, D3). In fact, the third Ig domain is encoded by a common or "constant" 5′ exon (a), which is spliced to either of two 3′ "variable" exons (b or c): The choice of the 3′ exon dictates the ligand-binding specificity of the receptor (Werner et al., 1992; Miki et al., 1991, 1992). For example, alternative splicing of FGFR2 creates two differentially responsive isoforms, KGFR/FGFR2b and FGFR2c (Miki et al., 1992). Whereas the KGFR preferentially binds acidic FGF (aFGF), FGF3, KGF, and FGF10, FGFR2c binds aFGF, basic FGF (bFGF), FGF4, FGF5, FGF6, and FGF9 (Bottaro et al., 1990; Ornitz et al., 1996). In fact, all receptor members encoded by the FGFR gene family share overlapping recognition and redundant specificity (Givol and Yayon, 1992; Johnson and Williams, 1993; Mason, 1994). Given that multiple ligand and receptor isoforms exist, the diversity of FGFR ligand-binding specificity presents both a large combinatorial set of possible interactions, and a high degree of functional redundancy. This functional redundancy often makes obtaining interpretable results from knock-out mice problematic (see below). Finally, it is notable that the biological functions of FGFs also depend on interactions with heparins (or heparan sulfate proteoglycans) on the cell surface and in the extracellular matrix (Mason, 1994; LaRochelle et al., 1999).

In our studies of the KGFR-Ig fusion protein, the chimera initially provided a framework for dissection of its aFGF and KGF binding domains (Cheon et al., 1994). Prior analysis of a series of naturally occurring FGFR alternative products demonstrated that Ig-like domain 1 appeared dispensable for ligand-binding (Miki et al., 1991). Scatchard analysis revealed that the chimera containing KGFR Ig-like domains D2 and D3 bound KGF and aFGF at high affinities comparable with the native receptor. Most interesting, individual Ig-like domain chimeras demonstrated marked specificity in their ligand interactions. D2-Fc bound aFGF at high affinity, whereas it did not interact in any detectable way with KGF. Conversely, D3-Fc bound KGF at high affinity but exhibited no detectable interaction with aFGF. Selective ligand-binding properties were confirmed by the specific neutralization of aFGF or KGF mitogenic activity using D2 or D3 Fc, respectively. Our results demonstrated that the high-affinity binding sites for aFGF and KGF reside within individual KGFR Ig-like domains, D2 and D3, respectively. All of these data suggest that in contrast to the PDGFR, individual KGFR Ig-like domains were able to specify high-affinity ligand-binding.

The biologic activity of the KGFR is further illustrated in our study comparing dnFGFR-Fc and dnFGFR-Tm, an engineered KGFR/FGFR2b mutant that lacked the cytosolic kinase domain but retained the transmembrane domain, anchoring it to the cell membrane (Figs. 9.1 and 9.2A). To test and compare their biologic function in vitro, the two receptor mutants were overexpressed by using the mouse metallothionein (MT) I gene promoter (MMTneo vector) in Balb/MK cells, an epidermal cell line that expresses the native KGFR (Bottaro et al., 1990). Both dominant negative receptor forms selectively reduced the mitogenic activity of aFGF and KGF (Fig. 9.2B), but not that of EGF, HGF or IGF-1, indicating that they were equally effective dominant negative receptor forms in vitro.

The ligand binding specificity of the dnFGFR-Fc was further illustrated in a mitogenic assay using serum-starved NIH 3T3 cells treated overnight with increasing concentrations of purified dnFGFR-Fc preincubated with 1 ng/ml aFGF, bFGF, FGF3, or FGF4. As shown in Figure 9.2C, dnFGFR-Fc inhibited the mitogenic response to aFGF completely and to FGF3 less effectively, but it did not affect bFGF or FGF4 activity. To determine that the dnFGFR-Fc mutant was actually blocking signaling through endogenous FGFRs, quiescent NIH 3T3 mouse fibroblasts were exposed to aFGF or bFGF preincubated with increasing amounts of the purified soluble dominant negative receptor. Endogenous FGFR1 phosphotyrosine content was quantitated. The dnFGFR-Fc mutant blocked phosphorylation of FGFR1 in response to aFGF, but not bFGF, indicating that the soluble chimera functions as a dominant negative inhibitor by interfering with specific FGF-induced FGFR signaling (Fig. 9.2D).

Soluble Dominant Negative Receptors in Transgenic Mice: The dnFGFR-Fc Chimera as a Potent In Vivo Antagonist of Receptor Signaling

A number of soluble receptors or receptor-Ig fusion proteins have now been successfully utilized as dominant negative agents in vivo. For example: a soluble insulin-like growth factor II/mannose 6-phosphate receptor can inhibit the growth of specific transgenic mouse organs; soluble tumor necrosis factor (TNF) receptor-Fc fusion proteins have been used to demonstrate roles for TNF in lymphoid development, lung inflammation, and salivary autoimmune disease (Ettinger et al., 1998; Smith et al., 1998; Hunger et al., 1996); in vivo expression of a soluble lymphotoxin-β receptor-Fc fusion product disrupted development of the spleen and Peyer's patches (Ettinger et al., 1996); and TIE2-Fc transgenic mice exhibited severely impaired definitive hematopoiesis (Takakura et al., 1998). We now focus on the practical application of soluble-Fc receptor chimeras toward elucidating the in vivo role of FGFR signaling in embryogenesis (described in Celli et al., 1998).

Fibroblast Growth Factors, Their Receptors, and Development. FGFs belong to a superfamily of signaling molecules that regulate a wide variety of functions, including development, angiogenesis, wound healing, and malignant transformation (reviewed by Burgess and Maciag, 1989; Basilico and Moscatelli, 1992; Mason, 1994; Yamaguchi and Rossant, 1995). FGFs are known to be potent mitogens, not only for cells of mesodermal and neuroectodermal origin, but for ectodermal and endodermal derivatives as well. However, their functions are not limited to induction of proliferation. FGFs have been shown in vitro to modulate motility, differentiation, neurite extension, and cell survival; whereas in vivo, expression of some FGFs in adult tissues implicates their activities in normal homeostasis. Interest in this receptor family was intensified by the recent discovery that mutations in FGFRs are associated with hereditary autosomal dominant human skeletal disorders involving craniofacial and limb anomalies, including Apert, Pfeiffer, Jackson-Weiss, Crouzon, and Beare-Stevenson cutis gyrata syndromes, and achondroplasia and thanatophoric dysplasias (reviewed by Muenke and Schell, 1995; Webster and Donoghue, 1997). The alteration of receptor structure and loss of requirement for ligand by these mutations is thought to result in the constitutive activation of FGFRs.

A striking characteristic of the FGF family is that their members are widely—and some exclusively—expressed during embryogenesis. In fact, on the basis of their embryonic expression patterns, FGFs and their receptors have long been implicated in vertebrate development (reviewed in Rubin et al., 1995; Yamaguchi and Rossant, 1995; Martin, 1998; Szebenyi and Fallon, 1999). Some years ago, targeted gene disruption of FGF4 and FGFR1 confirmed that FGF signaling was required for normal blastocyst growth, primary mesoderm induction, and early pattern formation in mouse (Yamaguchi et al., 1994; Deng et al., 1997; Feldman et al., 1995). Indication that FGFs act as mediators of mesenchymal-epithelial cell interactions during the formation of structures such as limb and lung has come from use of FGF implants, organ culture, antisense oligonucleotides, and neutralizing antibodies (see reviews by Mason, 1994; Yamaguchi and Rossant, 1995; Martin, 1998; Szebenyi and Fallon, 1999). Nevertheless, obtaining more definitive genetic data for a role in organogenesis from gene-targeting approaches had been problematic because of early embryonic lethality prior to full organ induction, as with homozygous null mutations in FGFR1, FGF4, FGF8, and FGFR2 (Yamaguchi et al., 1994; Deng et al., 1996, 1997; Feldman et al., 1995; Meyers et al., 1998), or because of functional redundancy within both the FGF and FGFR families. Such redundancy can cause incomplete phenotypic expression in null embryos, and an underestimation of the role of specific ligands and/or receptors in embryonic development. However, gene targeting technology has successfully shown the following: FGF3 inactivation perturbs normal development of the ear and tail (Mansour et al., 1993); FGFR3 null mutants exhibit overgrown long bones and abnormal curvature of the spine and tail (Deng et al., 1996); aberrant hair growth and development results from loss of FGF5 or KGF function (Hébert et al., 1994); FGF6 inactivation disrupts adult skeletal muscle regeneration (Floss et al., 1997); bFGF null mice show non-lethal defects in blood-pressure regulation and neuronal development (Zhou et al., 1998; Ortega et al., 1998; Dono et al., 1998); and FGF10 null fetuses develop all the way to birth but then die lacking both lungs and limbs (Sekine et al., 1999).

Creation of Dominant Negative FGFR Transgenic Mice.   Dominant negative transgenic strategies have been used with mixed success to overcome functional redundancies and early embryonic lethality associated with the FGFR superfamily. Truncated membrane-bound FGFRs that lack a functional tyrosine kinase domain have been shown in vitro to disrupt FGFR signaling of multiple receptor isoforms by competing for ligand-binding and by forming inactive heterodimers with endogenous FGFRs (Ueno et al., 1992). In vivo, this dominant negative approach has been used to show that FGFs are required for normal development in *Xenopus* (Amaya et al., 1991), and in mouse (Werner et al., 1993, 1994; Peters et al., 1994; Robinson et al., 1995; Chow et al., 1995; Campochiaro et al., 1996; Saffell et al., 1997; Jackson et al., 1997; Chai et al., 1998). However, the efficacy of defective membrane-bound receptors is limited by the spatial and temporal expression properties of the transcriptional promoter directing expression of the transgene to specific cell types, and by the need to be greatly overexpressed to effectively compete for ligand-binding with native receptors at the cell surface. We decided to compare the inhibitory effects of the kinase-deficient, membrane-bound dnFGFR-Tm mutant with that of the soluble dnFGFR-Fc and found that the secreted chimera was far superior to the membrane-bound form at perturbing development, and much more effective as a dominant negative agent in vivo.

The mouse MT promoter, flanked by its locus control regions (Palmiter et al., 1993), was used to achieve broad reproducible expression in transgenic mice of either the membrane-tethered dnFGFR-Tm mutant, or the soluble dnFGFR-Fc chimera (Figs. 9.1 and 9.2A). The MT-dnFGFR-Tm DNA construct was microinjected into single-cell mouse zygotes, and surviving embryos were transferred into surrogate mothers to complete development. Three viable founder transgenic animals were identified out of 22 possible founders screened by polymerase chain reactions (PCR). With Northern blot hybridization, these mice were characterized by the presence of abundant transgenic RNA transcripts in many adult tissues, including liver and kidney, and in midgestation embryos in two of the three lines. Despite this level of transgene expression, significant phenotypic effects in these mice were not apparent.

In contrast, attempts to establish MT-dnFGFR-Fc transgenic lines were completely unsuccessful (0/89 possible founders screened by PCR), raising the possibility that expression of the soluble chimera was incompatible with normal mouse development. Therefore, we decided to employ a "transient" transgenic approach to determine the consequences of expression of this soluble chimera. Mouse zygotes microinjected with the MT-dnFGFR-Fc transgene were allowed to develop to 18.5 days post-coitum (or E18.5) in the surrogate mother, and the resulting fetuses harvested for visual, histological, and molecular inspection. Fifteen percent of the viable fetuses were found to be grossly abnormal, and all were shown by PCR analysis to carry the MT-dnFGFR-Fc transgene. Abnormal transgenic fetuses were small and exhibited limb truncations of varying severity, ranging from shortening of the most distal elements to complete loss of the appendages (Fig. 9.3A). Typically, limb phenotypes were accompanied by craniofacial anomalies, curly tails, thin featureless skin, and open eyes. In situ hybridization was used to show that dnFGFR-Fc transcripts were broadly expressed in the embryo. Immunohistochemical analysis using an anti-mouse IgG (Fc) antibody revealed that the soluble chimeric protein was localized to the extracellular milieu in various embryonic tissues.

Western blot analysis using the same anti-mouse IgG antibody demonstrated that the variability in penetrance and expressivity of the phenotypes observed in these "transient" transgenic embryos could be explained by differential transgene expression, resulting from the influence of different genomic sequences flanking the site of integration of each individual transgene (Celli et al., 1998). The possibility that the observed phenotypes were due to expression of the Fc portion of the chimera independent of the FGFR domain was discounted through analysis of additional transgenic mice expressing different Fc chimeras.

The Soluble FGFR-Fc Chimera Potently Disrupts Organogenesis. Ninety percent of E18.5 MT-dnFGFR-Fc embryos displayed gross abnormal limb and craniofacial features (Fig. 9.3A) reminiscent of human dominant skeletal disorders associated with mutations in FGFR genes (see above). Therefore, it was of great interest to examine appendicular and craniofacial skeletal development in these transgenic embryos. In dnFGFR-Fc transgenic embryos, truncations of fore-limb and hind-limb skeleton occurred at different levels along the proximodistal limb axis, presumably reflecting the time of onset of transgene expression in a given embryo (Fig. 9.3B–9.3D). Analysis of the skull of E18.5 transgenic embryos revealed a widely cleft palate, a reduced maxillary bone, and an absent otic capsule (Fig. 9.3G–9.3J). Therefore, the presence of the dnFGFR-Fc effectively disrupted development of many skeletal structures, demonstrating abnormalities in FGFR-associated human syndromes.

Histopathological analysis of E18.5 embryos expressing the soluble dnFGFR-Fc mutant revealed abnormal development in many organs previously demonstrated to express FGF receptors (see Table 9.1, and Celli et al., 1998). Typically, the lungs were absent, and the respiratory system consisted only of the trachea and rudimentary primary bronchi that ended abruptly without further branching. In the head/neck region, tooth buds did not form; the thymus was small, hypoplastic, and lacked a defined medulla and cortex; the thyroid, salivary, and pharyngeal serous glands were missing; the eye was relatively small and never formed eyelids; the anterior pituitary was absent; and only a rudiment of the inner ear was present. Kidneys were either small and highly abnormal or they failed to develop altogether. Of the gastrointestinal organs, the pancreas was reduced to a few, morphologically aberrant acinar cells and no

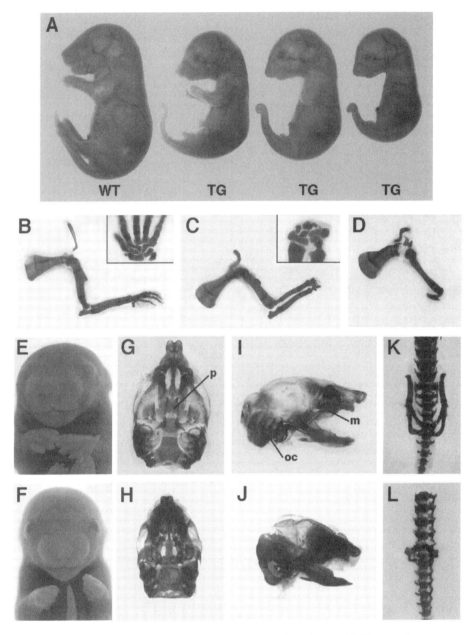

Figure 9.3. Skeletal phenotypes in dnFGFR-Fc transgenic embryos. (**A**) Transgenic (TG) E18.5 embryos with increasingly severe limb phenotypes and a wild-type (WT) littermate. Normal fore-limb skeleton (**B**) stained with Alizarin red and Alcian blue compared with those of transgenic (**C,D**) embryos shown in (**A**). Inserts show normal carpal bones at distal end of truncated limb compared with WT. Frontal view of normal (**E**) and transgenic (**F**) littermates. Note the open eyes and the loss of toenails. Ventral (**G, H**) and side (**I, J**) views of skulls from control (**G, I**) and transgenic (**H, J**) embryos. Palantine bone (p), otic capsule (oc), and maxillary bone (m) are indicated. (**L**) Rudimentary pelvis from mutant embryo shown on far right in (**A**), compared with that of a normal littermate (**K**). Taken with permission from Celli et al., 1998. (See color plates.)

Table 9.1. Incidence of histological abnormalities in E18.5 embryos expressing soluble FGFR2-Ig Chimera[a]

	Lung	Anterior Thyroid	Pituitary	Tooth Buds	Salivary Glands	Kidney	Hair Follicles	Inner Ear	Thymus	Glandular Stomach	Pancreas
No. Embryos Examined	8	5	3	6	5	8	8	8	8	8	8
Dysgenesis[b]	–	–	–	–	–	3	4	8	8	8	8
Agenesis[b]	8	5	3	6	5	5	4	–	–	–	–

[a] Taken from Celli et al., 1998.

[b] Dysgenesis is defined as malformation or reduction in number or size, and agenesis as absence of organ.

detectable islets; the liver was small and exhibited increased red blood cell hematopoiesis; and the glandular portion of the stomach was reduced or missing. In contrast, the small and large intestine and the reproductive organs appeared to be overtly normal. The skin demonstrated a partial or complete loss of hair follicles. Moreover, detailed immunohistochemical analysis of proliferative, apoptotic, and differentiation markers revealed that the normal program of differentiation had been significantly altered in the transgenic skin.

Analysis of Inductive Interactions in FGFR Signaling-Deficient Embryos.
Application of these Fc chimeras also permits molecular dissection of those FGFR-based mesenchymal-epithelial inductive interactions responsible for directing normal organogenesis. Midgestational transgenic embryos were subjected to in situ hybridization with a variety of diagnostic molecular markers (see Celli et al., 1998, for details). As one example, analysis of limb-bud development is shown. In the developing limb, expression of the homeobox gene *Msx-1*, but not *Msx-2*, in the distal mesenchyme requires the presence of a functional apical ectodermal ridge, or AER (Wang and Sassoon, 1995, and references therein). The limb-bud rudiment in a severely affected transgenic E12.5 embryo was identified by the presence of an anterior *Msx-2* domain, but transcripts for *Msx-1* were undetectable in mesenchyme and ectoderm, indicating that a functional as well as morphological AER was lost by E12.5, or had never formed (Figs. 9.4A and 9.4B). In confirmation, immunohistochemical detection of nuclear incorporation of 5-bromo-2'-deoxyuridine (Brdu) showed that distal mesenchymal proliferation was absent.

The kidney represents an excellent second example. Development of the definitive kidney requires reciprocal inductive interactions between the metanephric blastema and the ureteric bud. During the resulting mesenchymal-epithelial conversion, the paired box gene *Pax-2* is expressed in condensed mesenchyme apparently in response to early epithelial signals (Figs. 9.4C and 9.4D) (Eccles et al., 1995). However, *Pax-2* transcripts were not detectable in the rudimentary blastema of overtly affected E12.5 transgenic embryos, indicating that expression of the dnFGFR-Fc chimera disrupted reciprocal inductive signals that normally induce ureteric-bud outgrowth and/or mesenchymal-epithelial conversion (Figs. 9.4E and 9.4F).

Finally, we turn to the anterior pituitary. The pituitary is formed from the mutual interactions between the ectodermal primordium of the anterior and intermediate lobes of the pituitary, or Rathke's pouch, and the floorplate neuroectoderm. Once Rathke's pouch has formed, its epithelial lining proliferates and differentiates into distinct cell lineages (Rhodes et al., 1994). By E12.5, dnFGFR-Fc transgenic embryos exhibited the infundibular recess, but not Rathke's pouch (compare Figs. 9.4G and 9.4I). The absence of Rathke's pouch was confirmed by the loss of expression of *Lim 3*, a LIM homeobox gene involved in the development of the anterior and intermediate lobes of the pituitary (Fig. 9.4J). However, expression of Sonic hedgehog (*Shh*), a segment polarity gene expressed in multiple embryonic tissues—including floorplate neuroectoderm—was detected at the base of the diencephalon in transgenic embryos (data not shown), indicating that although FGFR signaling is required for proper development of the anterior pituitary, it acts downstream of, or in parallel to, *Shh*.

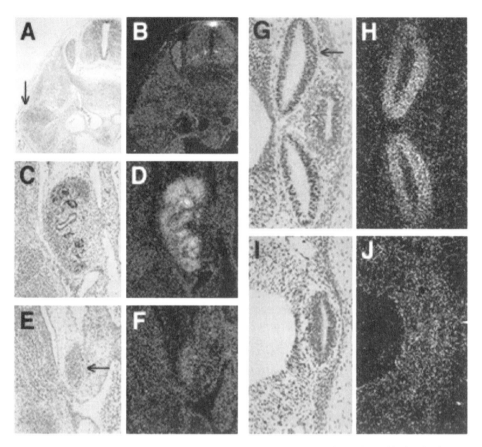

Figure 9.4. Diagnostic molecular markers demonstrate disruption of inductive interactions in transgenic embryos expressing the soluble dnFGFR-Fc chimera. Analysis of RNA transcripts was by in situ hybridization using ^{33}P-labeled antisense riboprobes. Shown are wild-type (**C, D, G, H**) and transgenic (**A, B, E, F, I, J**) E12.5 embryos. Dark-field pictures are indicated by white letters. In situ riboprobes: *Msx-1* (**A, B**), *Pax-2* (**C-F**) and *Lim 3* (**G-J**). Note that the residual transgenic limb bud (**A/B**, arrow) fails to express *Msx-1*, whereas *Pax-2* expression is absent from the rudimentary metanephric transgenic blastema (**E/F**, arrow). Also, note that the presumptive anterior pituitary, Rathke's pouch marked by expression of *Lim 3* (**G/H**, arrow), does not form in the transgenic embryos (**I/J**). Magnifications: (**A, B**) 50X; (**C-F**) 100X; (**G-J**) 200X. Taken in part, and with permission, from Celli et al., 1998.

Using this dominant negative approach, we have shown that FGFR signaling is essential for inductive interactions and patterning associated with the formation of many organs and structures in the murine embryo. Together, these data demonstrate that the soluble FGFR-Fc chimera is an extremely effective agent for uncovering the in vivo roles of FGFs and their receptors. This finding is especially obvious when compared with the lack of phenotypic consequences observed in transgenic embryos expressing the membrane-tethered, kinase-deficient FGFR mutant. We believe that the soluble chimera is vastly superior because, as a secreted dimer, it can bind and effectively inactivate FGF ligands before they have the opportunity to interact with native, cell-bound FGFRs. In contrast, the membrane-tethered mutant must directly compete at the cell surface for ligand with native FGFRs, or inactivate such wild-type receptors through heterodimerization (see Fig. 9.1).

Future Perspectives on Soluble Dominant Negative Agents

Chimeric receptor-Ig fusion proteins have proven valuable in studying protein-protein interactions, probing biologic function, and isolating orphan ligands, and these proteins may possess diagnostic as well as therapeutic applications. Their potency as genetic dominant negative agents make them ideally suited to dissect receptor function in the context of the whole animal. By exploiting specific ligand recognition and promoter targeting, β PDGFR-Fc and KGFR-Fc chimeras are proving useful for analysis of the distribution and expression levels of individual ligands in tissues. Moreover, the ability to identify and separate receptor sites involved in high-affinity interactions with different related ligands could provide a powerful approach toward in vitro screening of specific agonists and antagonists of FGFR function. In vivo studies using a CD4-Fc chimeric molecule to block HIV infectivity demonstrated that the Fc portion increased the plasma half-life of soluble CD4 and possessed the added advantage of IgG Fc-mediated placental transfer (Capon et al., 1989). Therapeutically, TGFβ receptor-Fc chimeras suppressed extracellular matrix accumulation in experimentally induced glomerulonephritis in the rat (Isaka et al., 1999). In humans, TNFR-Fc chimeras have been used to reduce the symptoms associated with OKT3 acute clinical syndrome after induction of renal allograft recipients (Novak et al., 1998).

Our laboratory has also developed growth factor-Ig chimeras that may be generally applicable in the development of highly specific probes for cell-surface receptors and in targeting molecules to intracellular sites of specific cell populations (LaRochelle et al., 1995). These studies were undertaken because the nearly identical nature of the various FGFR2 isoforms has made it difficult to develop specific immunochemical probes. Chimeric KGF-Fc fusion proteins defined the distribution of specific receptors in histologic specimens and may prove useful in diagnosis of malignancies, such as cervical carcinoma (Diez de Medina et al., 1997). The ability of growth factor-Ig fusion proteins to cause functional receptor activation suggests normal receptor processing associated with down-regulation and internalization. If so, such chimeric ligands may provide ideal vehicles for delivering toxins and other therapeutic modalities to intracellular sites in specific target cell populations.

REFERENCES

Amaya E, Musci TJ, and Kirschner MW (1991): Expression of a dominant negative mutant of the FGF receptor disrupts mesoderm formation in *Xenopus* embryos. Cell 66:257–270.

Amundadottir LT, Merlino G, Dickson RB (1996): Transgenic mouse models of breast cancer. Breast Canc 39:119–135.

Basilico C, Moscatelli D (1992): The FGF family of growth factors and oncogenes. Adv Canc R 59:115–165..

Bennett BD, Bennett GL, Vitangcol RV, Jewett JRS, Burnier J, Henzel W, et al. (1991): Extracellular domain-IgG fusion proteins for three human natriuretic peptide receptors. J Biol Chem 266:23060–23067.

Bottaro DP, Rubin JS, Ron D, Finch PW, Florio C, Aaronson SA (1990): Characterization of the receptor for keratinocyte growth factor. Evidence for multiple fibroblast growth factor receptors. J Biol Chem 265:12767–12770.

Bottinger EP, Jakubczak JL, Roberts ISD, Mumy M, Hemmati P, Bagnall K, et al. (1997): Expression of a dominant-negative mutant TGF-b type II receptor in transgenic mice reveals essential roles for TGF-b in regulation of growth and differentiation in the exocrine pancreas. EMBO J 18:2621–2633.

Burgess WH, Maciag T (1989): The heparin-binding (fibroblast) growth factor family of proteins. Ann Rev Bioch 58:575–606.

Burton DR (1985): Immunoglobulin G: functional sites. Mol Immunol 22:161–206.

Capon DJ, Chamow SM, Mordenti J, Marsters SA, Gregory T, Mitsuya H, et al. (1989): Designing CD4 immunoadhesins for AIDS therapy. Nature 337:525–531.

Campochiaro PA, Chang M, Ohsato M, Vinores SA, Nie Z, Hjelmeland L, et al. (1996): Retinal degeneration in transgenic mice with photoreceptor-specific expression of a dominant negative fibroblast growth factor receptor. J Neurosc 16:1679–1688.

Celli G, LaRochelle WJ, Mackem S, Sharp R, Merlino G (1998): Soluble dominant-negative receptor uncovers essential roles for fibroblast growth factors in multi-organ induction and patterning. EMBO J 17:1642–1655.

Chai N, Patel Y, Jacobson K, McMahon J, McMahon A, Rappolee DA (1998): FGF is an essential regulator of the fifth cell division in preimplantation mouse embryos. Dev Bio 198:105–115.

Chang P-Y, Goodyear LJ, Benecke H, Markuns JS, Moller DE (1995): Impaired insulin signaling in skeletal muscle from transgenic mice expressing kinase-deficient insulin receptors. J Biol Chem 270:12593–12600.

Cheon H-G, LaRochelle WJ, Bottaro DP, Burgess WH, Aaronson SA (1994): High-affinity binding sites for related fibroblast growth factor ligands reside within different receptor immunoglobulin-like domains. P NAS US 91:989–993

Chin L, Merlino G, DePinho RA (1998): Malignant melanoma: modern black plague and genetic black box. Gene Dev 12:3467–3481.

Chow RL, Roux GD, Roghani M, Falmer MA, Rifkin DB, Moscatelli DA, et al. (1995): FGF suppresses apoptosis and induces differentiation of fibre cells in the mouse lens. Development 121:4383–4393.

Damm K, Heyman RA, Umesono K, Evans RM (1993): Functional inhibition of retinoic acid response by dominant negative retinoic acid receptor mutants. P NAS US 90:2989–2993.

Davies DR, Sheriff S, Padlan EA (1988): Antibody-Antigen complexes. J Biol Chem 263:10541–10544.

Deng C, Wynshaw-Boris A, Zhou F, Kuo A, Leder P (1996): Fibroblast growth factor receptor 3 is a negative regulator of bone growth. Cell 84:911–921.

Deng C, Bedford M, Li C, Xu X, Yang Y, Dunmore J, et al. (1997): Fibroblast growth factor receptor 1 (FGFR1) is essential for normal neural tube and limb development. Dev Bio 185:42–54.

Diez da Medina S, Chopin D, El Marjou A, Delouvee A, LaRochelle WJ, Hoznek A, et al. (1997): Decreased expression of keratinocyte growth factor receptor in a subset of human transitional cell bladder carcinomas. Oncogene 14:323–330.

Dono R, Texido G, Dussel R, Ehmke H, Zeller R (1998): Impaired cerebral cortex development and blood pressure regulation in FGF-2-deficient mice. EMBO J 17:4213–4225.

Dumont DJ, Gradwohl G, Fong G-H, Puri MC, Gertsenstein M, Auerbach A, et al. (1994): Dominant-negative and targeted null mutations in the endothelial receptor tyrosine kinase, tek, reveal a critical role in vasculogenesis of the embryo. Gene Dev 8:1897–1909.

Duncan AR, Woof JM, Partridge LJ, Burton DR, Winter G (1988): Localization of the binding site for the high-affinity Fc receptor on IgG. Nature 332:563–564.

Eccles MR, Yun K, Reeve AE, Fidler AE (1995): Comparative in situ hybridization analysis of PAX2, PAX8, and WT1 gene transcription in human fetal kidney and Wilms' tumors. Am J Path 146:40–45.

Eilat D, Kikuchi GE, Coligan JE, Shevach EM (1992): Secretion of a soluble, chimeric gd T-cell receptor-immunoglobulin heterodimer. P NAS US 89:6871–6875.

Ettinger R, Browning JL, Michie SA, van Ewijk W, McDevitt HO (1996): Disrupted splenic architecture, but normal lymph node development in mice expressing a soluble lymphotoxin-b receptor-IgG1 fusion protein. P NAS US 93:13102–13107.

Ettinger R, Mebius R, Browning JL, Michie SA, van Tuijl S, Kraal G, et al. (1998): Effects of tumor necrosis factor and lymphotoxin on peripheral lymphoid tissue development. Int Immunol 10:727–741.

Ey PL, Prowse SJ, Jenkin CR (1978): Isolation of pure IgG1, IgG2a, and IgG2b immunoglobulins from mouse serum using protein A-Sepharose. Immunochemistry 15:429–436.

Feldman B, Poueymirou W, Papaioannou VE, DeChiara TM, Goldfarb M (1995): Requirement of FGF4 for postimplantation mouse development. Science 267:246–249.

Feng X, Peng ZH, Di W, Li XY, Rochette-Egly C, Chambon P, et al. (1997): Suprabasal expression of a dominant-negative RXR alpha mutant in transgenic mouse epidermis impairs regulation of gene transcription and basal keratinocyte proliferation by RAR-selective retinoids. Gene Dev 11:59–71.

Floss T, Arnold HH, Braun T (1997): A role for FGF6 in skeletal muscle regeneration. Gene Dev 11:2040–2051.

Givol D, Yayon A (1992): Complexity of FGF receptors: genetic basis for structural diversity and functional specificity. FASEB J 6:3362–3369.

Gascoigne NJR, Goodnow C, Dudzik KI, Oi VT, Davis, MM (1987): Secretion of a chimeric T-cell receptor-immunoglobulin protein. P NAS US 84:2936–2940.

Hébert JM, Rosenquist T, Götz J, Martin GR (1994): FGF5 as a regulator of the hair growth cycle: evidence from targeted and spontaneous mutations. Cell 78:1017–1025.

Heidaran MA, Yu J-C, Jensen RA, Matsui T, Aaronson SA (1990): Chimeric a and b-platelet-derived growth factor receptors define three immunoglobulin-like domains of the a-PDGF receptor that determine PDGF AA binding specificity. J Biol Chem 265:18741–18744.

Heidaran MA, Mahadevan D, LaRochelle WJ (1995): b PDGFR-IgG chimera demonstrates that human b PDGFR Ig-like domains 1 to 3 are sufficient for high-affinity PDGF BB binding. FASEB J 9:140–145.

Howard OMZ, Clouse KA, Smith C, Goodwin RG, Farrar WL (1993): Soluble tumor necrosis factor receptor: inhibition of human immunodeficiency virus activation. P NAS US 90:2335–2339.

Hunger RE, Muller S, Laissue JA, Hess MW, Carnaud C, Garcia I, et al. (1996): Inhibition of submandibular and lacrimal gland infiltration in non-obese diabetic mice by transgenic expression of soluble TNF-receptor p55. J Clin Inv 15:954–961.

Imakado S, Bickenbach JR, Bundman DS, Rothnagel JA, Attar PS, Wang XJ, et al. (1995): Targeting expression of a dominant negative retinoic acid receptor mutant in the epidermis of transgenic mice results in loss of barrier function. Gene Dev 9:317–329.

Isaka Y, Akagi Y, Ando Y, Tsujie M, Sudo T, Ohno N, et al. (1999): Gene therapy by transforming growth factor-beta receptor-IgG Fc chimera suppressed extracellular matrix accumulation in experimental glomerulonephritis. Kidney Int 55:465–475.

Jackson D, Bresnick J, Rosewell I, Crafton T, Poulson R, Stamp G, et al. (1997): Fibroblast growth factor receptor signaling has a role in lobuloalveolar development of the mammary gland. J Cell Sci 110:1261–1268.

Johnson DE, Williams LT (1993): Structural and functional diversity in the FGF receptor multigene family. Adv Canc R 60:1–41.

Kaplan JB, Sridharan L, Zaccardi JA, Dougher-Vermazen M, Terman BI (1997): Characterization of a soluble vascular endothelial growth factor receptor-immunoglobulin chimera. Grow Factor 14:243–256.

Kashles O, Yarden Y, Fischer R, Ullrich A, Schlessinger J (1991): A dominant negative mutations suppresses the function of normal epidermal growth factor receptors by heterodimerization. Mol Cell B 11:1454–1463.

LaRochelle WJ, Dirsch OR, Finch PW, Cheon H-G, May M, Marchese C, et al. (1995): Specific receptor detection by a functional Keratinocyte Growth Factor-Immunoglobulin chimera. J Cell Biol 129:357–366.

LaRochelle WJ, Sakaguchi K, Atabey N, Cheon HG, Takagi Y, Kinaia T, et al. (1999): Heparin proteoglycan modulates KGF signaling through interaction with both ligand and receptor. Biochemistry 38:1765–1771.

Letterio JJ, Bottinger EP (1998): TGF-beta knockout and dominant-negative receptor transgenic mice. Min Elect M Miner. Electrolyte Metab. 24:161–167.

Liu Y-C, Kawagishi M, Kameda R, Ohashi H (1993): Characterization of a fusion protein composed of the extracellular domain of c-kit and the Fc region of human IgG expressed in a baculovirus system. Bioc Biop R. 197:1094–1102.

Lyman SD, James L, Vanden Bos T, de Vries P, Brasel K, Gliniak B, et al. (1993): Molecular cloning of a ligand for the flt3/flk-2 tyrosine kinase receptor: a proliferative factor for primitive hematopoietic cells. Cell 75:1157–1167.

Lund J, Pound JD, Jones PT, Duncan AR, Bentley T, Goodall M, et al. (1992): Multiple binding sites on the C_H2 domain of IgG for mouse FcgRII. Mol Immunol 29:53–59.

Mahadevan D, Yu J-C, Saldanha JW, Thanki N, McPhie P, Uren A, et al. (1995): Structural role of extracellular domain 1 of the a platelet-derived growth factor receptor for PDGF A and PDGF B binding. J Biol Chem 270:27595–27600.

Mark MR, Lokker NA, Zioncheck TF, Luis EA, Godowski, PJ (1992): Expression and characterization of hepatocyte growth factor receptor-IgG fusion proteins. J Biol Chem 267:26166–26171.

Martin GR (1998): The roles of FGFs in early development of vertebrate limbs. Gene Dev 12:1571–1586.

Mason IJ (1994): The ins and outs of fibroblast growth factors. Cell 78:547–52.

Merlino G (1994): Transgenic mice as models for tumorigenesis. Canc Inv 12:203–213.

Meyers EN, Lewandoski M, Martin GR (1998): An FGF8 mutant allelic series generated by Cre- and Flp-mediated recombination. Nat Genet 18:136–141.

Miki T, Fleming TP, Bottaro DP, Rubin JS, Ron D, Aaronson SA (1991): Expression cDNA cloning of the KGF receptor by creation of a transforming autocrine loop. Science 251:72–75.

Miki T, Bottaro DP, Fleming TP, Smith CL, Burgess WH, Chan AM, et al. (1992): Determination of ligand-binding specificity by alternative splicing: two distinct growth factor receptors encoded by a single gene. P NAS US 89:246–250.

Millauer B, Shawver LK, Plate KH, Risau W, Ullrich A (1994): Glioblastoma growth inhibited in vivo by a dominant-negative Flk-1 mutant. Nature 10:576–579.

Muenke M, Schell U (1995): Fibroblast growth factor receptor mutations in human skeletal disorders. Trends Gen 11:308–313.

Murillas R, Larcher F, Conti CJ, Santos M, Ullrich A, Jorcano JL (1995): Expression of a dominant negative mutant of epidermal growth factor receptor in the epidermis of transgenic mice elicits striking alterations in hair follicle development and skin structure. EMBO J 14:5216–5223.

Novak EJ, Blosch CM, Perkins JD, Davis CL, Barr D, McVicar JP, et al. (1998): Recombinant human tumor necrosis factor Fc fusion protein therapy in kidney transplant recipients undergoing OKT3 induction therapy. Transplantation 66:1732–1735.

Ornitz DM, Xu J, Colvin JS, McEwen DG, MacArthur CA, Coulier F, et al. (1996): Receptor specificity of the fibroblast growth factor family. J Biol Chem 271:15292–15297.

Ortega S, Ittmann M, Tsang SH, Ehrlich M, Basilico C (1998): Neuronal defects and delayed wound healing in mice lacking fibroblast growth factor 2. P NAS US 12:5672–5677.

Palmiter RD, Sandgren EP, Koeller DM, Brinster RL (1993): Distal regulatory elements from the mouse metallothionein locus stimulate gene expression in transgenic mice. Mol Cell B 13:5266–5275.

Peters K, Werner S, Liao X, Wert S, Whitsett J, Williams L (1994): Targeted expression of a dominant negative FGF receptor blocks branching morphogenesis and epithelial differentiation of the mouse lung. EMBO J 13:3296–3301.

Ray P, Higgins KM, Tan JC, Chu TY, Yee NS, Nguyen H, et al. (1991): Ectopic expression of a c-kitW42 minigene in transgenic mice: recapitulation of W phenotypes and evidence for c-kit function in melanoblast progenitors. Gene Dev 5:2265–2273.

Rhodes SJ, DiMattia GE, Rosenfeld MG (1994): Transcriptional mechanisms in anterior pituitary cell differentiation. Cur Op Gen 4:709–717.

Robinson ML, MacMillan-Crow LA, Thompson JA, Overbeek PA (1995): Expression of a truncated FGF receptor results in defective lens development in transgenic mice. Development 121:3959–3967.

Rubin JS, Bottaro DP, Chedid M, Miki T, Ron D, Cheon H-G, et al. (1995): Keratinocyte Growth Factor. Cell Bio In 19:399–411.

Saffell JL, Williams EJ, Mason IJ, Walsh FS, Doherty P (1997): Expression of a dominant negative FGF receptor inhibits axonal growth and FGF receptor phosphorylation stimulated by CAMs. Neuron 18:231–242.

Saitou M, Sugai S, Tanaka T, Shimuouchi K, Fuchs E, Narumiya S, et al. (1995): Inhibition of skin development by targeted expression of a dominant-negative retinoic acid receptor. Nature 374:159–162.

Sekine K, Ohuchi H, Fujiwara M, Yamasaki M, Yoshizawa T, Sato T, et al. (1999): FGF10 is essential for limb and lung formation. Nat Genet 21:138–141.

Shohet JM, Pemberton P, Carroll, MC (1993): Identification of a major binding site for complement C3 on the IgG1 heavy chain. J Biol Chem 268:5866–5871.

Smith S, Skerrett SJ, Chi EY, Jones M, Mohler K, Wilson CB (1998): The locus of tumor necrosis factor-alpha action in lung inflammation. Am J Resp C 19:881–891.

Szebenyi G, Fallon JF (1999): Fibroblast growth factors as multifunctional signaling factors. Int Rev Cyt 185:45–106.

Takakura N, Huang XL, Naruse T, Hamaguchi I, Dumont DJ, Yancopoulos GD, et al. (1998): Critical role of the TIE2 endothelial cell receptor in the development of definitive hematopoiesis. Immunity 9:677–86.

Ueno H, Gunn M, Dell K, Tseng A, Williams LT (1992): A truncated form of fibroblast growth factor receptor 1 inhibits signal transduction by multiple types of fibroblast growth factor receptor. J Biol Chem 267:1470–1476.

Wang Y, Sassoon D (1995): Ectoderm-mesenchyme and mesenchyme-mesenchyme interactions regulate Msx1 expression and cellular differentiation in the murine limb bud. Develop Bio 168:374–382.

Webster MK, Donoghue DJ (1997): FGFR activation in skeletal disorders: too much of a good thing. Trends Gen 13:178–182.

Werner S, Duan DR, de Vries C, Peters KG, Johnso DE, Williams LT (1992): Differential splicing in the extracellular region of fibroblast growth factor receptor 1 generates receptor variants with different ligand binding specificities. Mol Cell B 12:82–88.

Werner S, Weinberg W, Liao X, Peters KG, Blessing M, Yuspa SH, et al. (1993): Targeted expression of a dominant-negative FGF receptor mutant in the epidermis of transgenic mice reveals a role of FGF in keratinocyte organization and differentiation. EMBO J 12:2635–2643.

Werner S, Smola H, Liao X, Longaker MT, Krieg T, Hofschneider PH, et al. (1994): The function of KGF in morphogenesis of epithelium and reepithelialization of wounds. Science 266:819–822.

Williams AF, Barclay AN (1988): The immunoglobulin superfamily—Domains for cell surface recognition. Ann R Immun 6:381–405.

Xie W, Paterson AJ, Chin E, Nabell LM, Kudlow JE (1997): Targeted expression of a dominant negative epidermal growth factor receptor in the mammary gland of transgenic mice inhibits pubertal mammary duct development. Mol Endocr 11:1766–1781.

Yamaguchi M, Nakamoto M, Honda H, Nakagawa T, Fujita H, Nakamura T, et al. (1998): Retardation of skeletal development and cervical abnormalities in transgenic mice expressing a dominant-negative retinoic acid receptor in chondrogenic cells. P NAS US 95:7491–7496.

Yamaguchi TP, Harpal K, Henkemeyer M, Rossant J (1994): Fgfr-1 is required for embryonic growth and mesodermal patterning during mouse gastrulation. Gene Dev 8:3032–3044.

Yamaguchi TP, Rossant J (1995): Fibroblast growth factors in mammalian development. Cur Op. Gen 5:485–491.

Yarden Y, Escobedo JA, Kuang W-J, Yang-Feng TL, Daniel TO, Tremble PM, et al. (1986): Structure of the receptor for platelet-derived growth factor helps define a family of closely related growth factor receptors. Nature 323:226–232.

Yu J-C, Mahadevan D, LaRochelle WJ, Pierce JH, Heidaran MA (1994): Structural coincidence of a PDGFR epitopes binding to PDGF AA and a potently neutralizing monoclonal antibody. J Biol Chem 269:10668–10674.

Zhou M, Sutliff RL, Paul RJ, Lorenz JN, Hoying JB, Haudenschild CC, et al. (1998): Fibroblast growth factor 2 control of vascular tone. Nat Med 4:201–207.

CHAPTER 10

TOXIN-MEDIATED TARGETED CELL ABLATION IN VIVO

JOHN DRAGO and JOHN Y. F. WONG

1. INTRODUCTION

The capacity to destroy specific cell populations or lineages is a powerful method for understanding signaling and cell-cell interactions during embryonic development and in the mature animal. The recent explosion in molecular biology and, in particular, transgenic and gene-targeting technology has provided investigators with a number of tools that enable the establishment of specific cellular lesions. The availability of cell-specific promoters that direct expression of toxic products to defined cell populations has been a major step forward. Temporal and spatial specificity has also been achieved by inserting genes encoding toxic products directly into well-characterized gene loci by the technique of homologous recombination in embryonic stem cells. Diphtheria toxin protein or Herpes Simplex Virus-1 thymidine kinase (HSV1-tk), in the presence of nucleoside analogues that inhibit DNA replication, have been the toxic mechanisms used in a large number of cell-ablation paradigms. Examples illustrating the use of each method will be presented highlighting the strengths and weaknesses.

2. PRODRUG CELL ABLATION PARADIGMS

Gene-directed enzyme prodrug paradigms have been widely used to establish models of human diseases and to study cell-cell interactions in vivo. The method involves the tissue-specific expression of a gene encoding an enzyme, which converts an inactive prodrug to a cell toxin. The assumption is that enzyme expression in itself is not damaging to cells and the prodrug is innocuous to tissues that lack the capacity to convert the prodrug into the active toxin. In permissive cells, the activated prodrug is an efficient cell killer and there are no ill effects on neighboring cells. This technique is powerful in that it gives the

Genetic Manipulation of Receptor Expression and Function,
Edited by Domenico Accili.
ISBN 0-471-35057-5   Copyright © 2000 Wiley-Liss, Inc.

investigator the advantage of determining the time of cell killing. In the commonly used experimental design, tissue-specific promoters are used to direct the expression of HSV1-tk. Phosphorylated products of prodrugs such as 1-(2-deoxy-2-fluoro-beta-D-arabino-furanosyl)-5-iodouracil (FIAU), acyclovir, or ganciclovir are thought to result in cell death by serving as defective nucleotide analogues, which are incorporated into cellular DNA during replication and cause chain termination.

The thymidine kinase cell obliteration system was first described in 1988 by Borrelli and colleagues (1988) and shortly thereafter applied in a transgenic paradigm to study the ontogeny of the endocrine cells of the anterior pituitary gland. Transgenic mice were generated carrying the HSV1-tk gene under the control of the rat growth-hormone promoter (Borrelli et al., 1989). Transgenic mice treated with FIAU were growth-retarded, and the anterior pituitary in treated animals was essentially free of both growth hormone and prolactin-producing cells, suggesting that there was a common progenitor cell for both cell lineages. The same general approach has also been used to address a number of developmental issues particularly during hemato/lymphopoiesis (Dzierzak et al., 1993; Minasi et al., 1993; Salomon et al., 1994). Thy-1 is a cell-surface marker that identifies mature T-cells, T-lymphoid precursors, and hematopoietic stem cells. The differentiation of Thy-1-positive cells at precise periods of in vivo development was studied using the Thy-1 promoter coupled to a HSV1-tk transgene. Near-total conditional ablation of maturing thymocytes was achieved using this strategy (Dzierzak et al., 1993). Ablation of interleukin-2 specific cells was achieved using an interleukin-2-specific promoter (Minasi et al., 1993), and dendritic cells that function as the principle antigen presenting cells pivotal in T-cell-mediated immune responses and in thymocyte differentiation were also ablated using Human Immunodeficiency virus regulatory domains thought to preferentially target transgene expression to antigen-presenting cells (Salomon et al., 1994).

There are two major limitations to the widespread applicability of the HSV-tk transgene approach. The first is the issue of the theoretical lack of toxicity of prodrug phosphorylation products in established post-mitotic cells expressing the HSV-tk transgene; the second is the problem of sterility in transgenic males (Al-Shawi et al., 1988; Salomon et al., 1995). With regard to the first problem, there are no studies in which the HSV-tk gene has been transactivated in adult neurons and efficient cell-killing achieved. There is, however, a model in which mitotically quiescent somatic cells have been efficiently ablated. Wallace and colleagues (1994) coupled HSV1-tk gene to a bovine thyroglobulin promoter and achieved efficient and rapid cell-killing of thyroid follicular cells after the administration of the prodrug ganciclovir. Cell ablation was shown to be due to apoptosis and to be independent of thyrocyte proliferation or nuclear DNA synthesis (Wallace et al., 1996). With regard to the second issue, the empirical observation that HSV1-tk transgenic males were invariably sterile was directly investigated by Braun and colleagues (1990) by expressing the HSV1-tk gene in spermatids using the protamine gene promoter, a promoter transactivated in post-meiotic germ cells. Braun and colleagues found that expression of the HSV1-tk in spermatids was sufficient to induce male sterility. In a subsequent study, the expression of the HSV1-tk gene was shown to be due to testis-specific high-level expression of short HSV1-tk transcripts initiated mainly upstream of a second internal ATG (Salomon et al., 1995). Transgenic mice lacking the cryptic initiation site located in DNA sequences upstream of the second ATG were generated in an attempt to obtain a suicide gene that did not cause male infertility. However, the mutant HSV1-tk gene behaved differently from the wild-type gene in that low-level expression did not confer sensitivity to ganciclovir and fertile transgenic males were generated at a very low frequency.

A novel prodrug system for achieving the inducible ablation of specific cell types in transgenic mice has been described by Clarke and colleagues (1997). It is based on the *E. Coli* enzyme nitroreductase, which coverts certain nitro compounds such as the antitumor drug 5-aziridin-1-yl-2-4-dinitrobenzamide, to powerful cell toxin. In this study, the gene encoding nitroreductase was expressed in the luminal cells of the mammary gland of transgenic mice using the ovine beta-lactoglobulin promoter. Treatment of transgenic mice with 5-aziridin-1-yl-2-4-dinitrobenzamide resulted in a rapid and selective killing of this population of cells, whereas the closely associated myoepithelial cells were unaffected.

3. CELL ABLATION BY TARGETED EXPRESSION OF DIPHTHERIA TOXIN

The diphtheria toxin consists of an A chain, which is known to inhibit protein synthesis by inhibition of ADP ribosylation of elongation factor-2 (Pappenheimer, 1977), and a B chain required for entry of the bioactive toxin into cells. Intracellular expression of gene sequence encoding the A chain is an efficient method of disrupting cell function and killing cells (Yamaizumi et al., 1978); when the A chain is expressed in cells in the absence of the B chain, direct toxin mediated bystander cell killing is not seen. A very large number of studies have used the diphtheria toxin gene to efficiently ablate cells in vivo and a very recent study has exploited a double transgenic/tetracycline paradigm to destroy cardiac myocytes (Lee et al., 1998), adding an exciting dimension to the expanding field of conditional tissue-specific cell-killing.

Deleting specific cell lineages by microinjection into fertilized eggs of a chimeric gene in which a cell-specific enhancer/promoter is used to drive the expression of the diphtheria toxin-A chain (DTA) gene was described independently in 1987 by two groups. The elastase I promoter/enhancer was used to generate mice lacking a normal pancreas due to the targeted expression of DTA to pancreatic acinar cells (Palmiter et al., 1987). Cell-killing was specific, as a small pancreatic rudiment containing islet and duct-like cells only was observed in transgenic mice. The mouse gamma 2-crystallin promoter fused to the coding region of the DTA gene (Breitman et al., 1987) was used to study lens development. Expressing transgenic mice and their transgenic offspring had small eyes containing lenses displaying considerable heterogeneity; some were structurally normal but reduced in size whereas, others were macroscopically abnormal and devoid of nuclear fiber cells. A number of subsequent studies used a similar tissue-specific promoter/enhance-DTA hybrid transgene paradigm to ablate a number of cell types, including pituitary growth-hormone producing cells (Behringer et al., 1988); alpha-crystallin-synthesizing lens cells

(Kaur et al., 1989); type II lung epithelial cells (Korfhagen et al., 1990); retinal photoreceptor cells (Lem et al., 1991); chondrocytes (Bruggeman et al., 1991); pituitary gonadotropes (Kendall et al., 1991); Schwann cells (Messing et al., 1992); vasoactive intestinal peptide positive cells (Gozes et al., 1993); brown adipose tissue (Lowell et al., 1993); cells transcribing the glucagon, insulin, or pancreatic polypeptide genes (Herrera et al., 1994); cerebellar Purkinje cells (Smeyne et al., 1995); rat osteocalcin-positive cells in bone and testis (Sato and Tada, 1995); granzyme A promoter-driven cytotoxic T-cells and CD8 T-cells (Aguila et al., 1995); gastric parietal cells (Li et al., 1996); and pituitary gonadotrope and thyrotrope cells (Burrows, et al., 1996).

Because of the extreme toxicity of wild-type diphtheria toxin to animal cells and the leakiness of many tissue-specific regulatory domains, an attenuated form of the diphtheria toxin-A chain gene (*tox-176*)—which is thought to be about 30-fold less potent than the wild-type gene—has been used by a number of investigators to ablate cells (Breitman et al., 1990; Ross et al., 1993; Raymond and Jackson, 1995; Burrows et al., 1996; Garabedian et al., 1997; Drago et al., 1998b). In one study, low- level *tox-176* expression was compatible with cell survival (Ross et al., 1993). Whole animal adiposity was reduced via targeted expression of *tox-176* to adipose tissue by using the adipocyte P2 promoter. Transgenic mice with high levels of toxin expression developed chylous ascites and died soon after birth. Transgenic mice expressing lower levels of the transgene had a normal amount of adipose tissue and survived to adulthood; however, they showed a complete resistance to chemically induced obesity suggesting that the *tox-176* gene product could specifically perturb cell function without inevitable cell death.

Transgene expression can have a devastating effect on phenotype and may make the establishment of a transgenic line impossible. This potential difficulty may be circumvented by the use of the *Cre*/LoxP system (Sauer and Henderson, 1988a; Sauer and Henderson, 1988b; Lakso et al., 1992; Sauer 1993). A typical binary transgenic system based on *Cre*-mediated DNA recombination to ablate cells was used to obtain skeletal muscle-deficient embryos by mating two phenotypically normal transgenic lines (Grieshammer et al., 1998). A complex transgene was constructed in which the time and site of DTA expression was controlled by the promoter of a muscle-specific gene myogenin, and expression of active DTA product occurred following *Cre* mediated in vivo recombination. A dormant DTA transgene was produced by disrupting the DTA coding sequence via insertion of the bacterial beta-galactosidase (*LacZ)* reporter gene between the myogenin promoter and the DTA gene. The *LacZ* reporter gene was in frame with the translation initiation codon of the DTA, contained a stop codon at its 3' end, and was flanked by LoxP sites, *Cre* recombinase-recognition sequences. The paradigm essentially involved *Cre*-mediated recombination of the two LoxP sites, to produce a single LoxP site with concomitant deletion of the *LacZ* and the generation of an in-frame DTA gene. The *Cre* transgene of the activator mice was under the control of the human beta-actin promoter. In beta-actin *Cre* embryos, *Cre* production is known to occur in all cells by the blastocyst stage of development (Lewandoski et al., 1997). In doubly transgenic embryos, skeletal muscles were eliminated secondary to the expression of the gene-encoding DTA as *Cre* recombinase removes LoxP bracketed *LacZ* encoding DNA. Complete muscle-cell ablation occurred by late embryonic development. Analysis of the consequences of muscle-cell ablation revealed that almost all motoneurons were lost by late embryogenesis.

In all the examples given above, transgenic lines have been generated by the conventional method of direct injection of DNA into the male pronucleus of a fertilized egg. The conventional transgenic approach has limitations because of the paucity of proven promoters, particularly neural-specific regulatory sequences, and the phenotype variability that occurs due to random transgene copy number and the site of integration in the genome. The gene-targeting approach (Capecchi, 1989) involves the precise introduction of a loss-of-function mutation in a gene of interest by homologous recombination in embryonic stem cells. Homologous recombination can also be used to introduce novel genes into specific gene loci (Le Mouellic et al., 1992; Drago et al., 1998b) and thereby avoid the loss of spatial and temporal fidelity in transgene expression sometimes seen when conventional transgenic approaches are used. We will now focus on a study in which specificity of *tox-176* transgene expression in D1 dopamine receptor (DIR) positive cells was achieved by a dormant gene knock-in paradigm.

Targeted Expression of a Toxin Gene to D1 Dopamine Receptor-Neurons By *Cre*-Mediated SiteSpecific Recombination

The neurotransmitter dopamine (DA) is thought to play a major role in a host of physiological neural processes (Jaber et al., 1996). In addition, Parkinson's disease (Hornykiewicz, 1966), Huntington's disease (Ginovart et al., 1997), drug-induced movement disorders (Kane et al., 1988), and several other common psychomotor disorders, including schizophrenia and Tourette's syndrome (Peterson, 1996), have been linked to alterations in DAergic signaling. Five subtypes of DA receptors have been cloned (Sibley and Monsma, 1992). The five receptors are subdivided into two classes, referred to as D1-like (D1 and D5, also called $D1_A$ and $D1_B$ respectively, in the rodent system), and D2-like (D2, D3, and D4) subclass (Sibley and Monsma, 1992). Individual receptor species mediate different effects by virtue of their intrinsic pharmacological binding properties for DA, downstream second messenger systems, and distribution within the nervous system. However, the specific in vivo role of each receptor has thus far been unclear because available ligands do not specifically distinguish between D1 and D5 DA receptors, or between D2, D3, and D4 DA receptors (Jaber et al., 1996). A number of DA receptor knockout mice have been generated, the analysis of which has provided some insight into the in vivo role of the individual DA receptor species (Drago et al., 1998a). The DA receptor expression profile (and therefore the function of cells within the CNS and the striatum in particular) is the subject of intense debate. A transgenic approach has now been used to dissect out the role of a genetically defined population of neurons in regulating movement. The cell-ablation paradigm complements the conventional DA receptor knockout approach of previous studies (Drago et al., 1994; Accili et al., 1996; Sibley et al., 1998) and provides a powerful model system to examine issues of fundamental biological and clinical relevance.

The DAergic system of the brain is prone to several degenerative conditions that disrupt normal basal ganglia function resulting in progressive motor impairment. Idiopathic Parkinson's disease is defined by loss of nigrostriatal DAergic neurons (Forno, 1982) and is characterized clinically by the presence of tremor, bradykinesia, rigidity and postural instability. Many brain diseases with similar clinical features also show loss of striatal neurons. Recent evidence suggests that striatal degeneration may not be confined to rare Parkinsonian syndromes (Steele et al., 1964; Lees 1987; Quinn 1994; Watts et al., 1994) but may be more common than generally appreciated due to misdiagnosis of cases as Idiopathic Parkinson's disease (Rajput et al., 1991; Hughes et al., 1992). In addition, hyperkinetic movement disorders such as Huntington's disease (Ferrante et al., 1985; Albin et al., 1989) and dystonia (Rothwell and Obeso, 1987; Waters et al., 1993) are also associated with striatal degeneration.

The striatum is a complex structure made up of projection neurons, interneurons, and afferent input from the brainstem and cerebral cortex (Graybiel, 1990; Gerfen, 1992). Medium-sized spiny projection neurons comprise ~90% of striatal neurons, with interneurons making up the remainder (Gerfen, 1992). Of the five cloned DA receptors (Sibley and Monsma, 1992), D1 and D2 DA (D2R) receptors are expressed at high levels in the adult striatum (Gerfen, 1992). D1R gene expression is detectable in the rat striatum from embryonic day 14 of development, with D1R mRNA being found at increasing levels in the caudate-putamen, accumbens nucleus, Islands of Calleja, and the olfactory tubercle during maturation (Caille et al., 1995). In contrast, D1R mRNA is not detectable in the dopaminergic neurons of the substantia nigra (Guennoun and Bloch, 1992). In situ hybridization studies suggested that D1Rs are preferentially expressed on substance P and dynorphin-positive striatal neurons, which project directly to the substantia nigra pars reticulata/entopeduncular complex (the striatonigral or direct pathway). Whereas enkephalin-positive/D2R neurons project to the same nuclear complex via the external segment of the globus pallidus and subthalamic nucleus (the striatopallidal or indirect pathway) (Gerfen et al., 1990). The validity of this dual pathway model has been challenged by recent studies reporting a substantial degree of D1 and D2 receptor co-localization on striatal projection neurons (Surmeier et al., 1992, 1993, 1996; Surmeier and Kitai, 1994).

Transgenic mice were generated using the strategy of insertion of a single copy of the *tox-176* gene (Maxwell et al., 1987) into the D1R locus by the method of homologous recombination in embryonic stem cells. In this model, the *tox-176* gene is in a functionally inactive state as the gene is located downstream from a LoxP flanked NEO/STOP cassette, which consists of the neomycin phosphotransferase gene (NEO) and a sequence known to inhibit downstream translation (STOP) (Lakso et al., 1992, 1996). Mutant mice producing the *tox-176* gene product selectively in D1R- positive cells were generated by mating conventional transgenic mice homozygous for the *Cre* transgene (Lakso et al., 1996) with phenotypically normal knock-in mice carrying the dormant *tox-176* gene (Drago et al., 1998b). The *Cre* transgene in this experiment is under the control of the Adenovirus EIIa promoter resulting in *Cre* expression in very early embryogenesis. Young post-natal day (P) 4 to P8 doubly transgenic mutant mice displayed a pleotropic neurological phenotype,

including slowing of locomotor activity (bradykinesia), twisting limb movements or dystonia, random intrusive body jerks usually resulting in falls, and
periodic breathing. The phenotype of the mutant mice was dramatic particularly when compared with the phenotype of other transgenic mice with null
mutations involving DA receptors or the DA transporter (DAT) (Drago et al.,
1998a). In a study in which tyrosine hydroxylase (TOH) was selectively inactivated in DAergic midbrain neurons, mutant mice were well in the first 2 weeks
of post-natal life (Zhou and Palmiter, 1995). However, all mice became brady-
kinetic, lost weight, and died by 4 weeks of age. The observations made in the
DA-deficient line suggest that factors other than simple perturbation of DA
neurotransmission underlies the dramatic phenotype of the *tox-176* mutant
mice. This hypothesis is supported by observations made on our D1R knockout
line and on data of an independently generated D1R knockout line produced by
Xu and colleagues (1994). Although changes in spontaneous behavior (Drago
et al., 1994; El-Ghundi et al., 1998: Clifford et al., 1998; Smith et al., 1998) and
baseline locomotor activity are described, the dramatic movement abnormalities and early post-natal death seen in *tox-176* mutant mice are not seen. This
finding suggests that loss of systems other than those mediated through D1Rs
are necessary to generate the repertoire of locomotor abnormalities seen in *tox-
176* mutants.

The brain of the *tox-176* mutants displayed features similar to those seen in
human neurodegenerative diseases, including apoptosis and reactive gliosis. As
expected from the high-level expression of D1Rs in the striatum and Islands of
Calleja seen in the normal mouse brain, there was a reduction in the volume of
the caudate/putamen, and the Islands of Calleja were not identified in doubly
transgenic mutant pups, whereas the cortex was of normal thickness. D1Rs
were not detected in mutants by in situ hybridization or ligand autoradiography although D2R mRNA and protein was present in the striatum. In addition,
substance P and dynorphin, neuropeptides known to be expressed in D1R-
positive striatonigral projection neurons were also not detected in mutant mice.
Enkephalin, a marker found in D2R-positive striatopallidal projection neurons
was expressed in the globus pallidus of the mutant brain (Drago et al., 1998b).
These observations confirmed, at least at the neurochemical level, that the destruction of the direct pathway was selective and complete. Given the postulated major role of target-derived growth factors in the establishment of neural
circuits (Barde, 1989), it was surprising that the nigrostriatal pathway in the
mutant pups was intact with a normal number of DAergic neurons in the substantia nigra and the ventral tegmental area, in addition to a normal pattern of
striatal DAT and TOH immunoreactivity. Quantitative analysis of striatal DAT
density using [^{3}H]-mazindol showed a moderate reduction, suggesting that a
degree of transneuronal downregulation occurs in these very young pups (P3-
P7). The number of healthy striatal neuropeptide Y-containing interneurons was
also substantially down-regulated in the mutant striatum. In contrast, vesicular
acetylcholine transporter (VAChT) immunohistochemistry, performed to identify nascent striatal cholinergic interneurons, revealed an increase in the number of VAChT-positive cells. Furthermore, upregulated striatal expression of
D2R mRNA, enkephalin mRNA, and VAChT immunoreactivity suggested that
apneic periods did not result in symptomatic striatal hypoxia sufficient to ex-

plain the apoptosis, reactive gliosis, or the pathophysiology of the locomotor phenotype. Down-regulated cortical $GABA_A$ and muscarinic receptor levels were also observed, in addition to subtle morphological changes in the NPY-expressing population of cortical neurons (Drago and Wong, unpublished observations).

Unlike younger pups, mutant pups examined in the third post-natal week had a hyperkinetic syndrome with gait abnormality, postural instability, and myoclonic jerks. The respiratory abnormality seen in young mutants was not seen in the cohort that was examined in the third post-natal week. Huntington's disease is typically a hyperkinetic syndrome associated with a spectrum of involuntary movements, including peripheral tic-like or choreoform movements and frequent falls due to myoclonic jerks in addition to cognitive and psychiatric manifestations. The clinical manifestation is largely determined by the age of onset, which is dependent on the number of CAG trinucleotide repeats (Nance, 1998). The Huntington's disease brain, although initially characterized as having selective loss of D2R/enkephalin-positive striatopallidal projection neurons, has more recently been found to have more extensive DA-receptor changes involving the D1-receptor system (Ginovart et al., 1997; Stewart et al., 1998) with down-regulated D1-class binding identified in both pre-symptomatic and symptomatic Huntington's- disease-gene-positive individuals (Andrews et al., 1998; Stewart et al., 1998). The phenotype of the older mutants more closely resembles the clinical Huntington's phenotype than the conventional transgenic mice generated by injection of the 5' fragment of an expanded Huntington's gene (Davies et al., 1997), the latter mice having a predominantly tremulous/seizures-disorder phenotype. Analysis of 3-week *tox-176* mutants shows that loss of the D1R positive cohort of cells is indeed sufficient to generate an age-dependent hyperkinetic/movement disorder phenotype. Striatal D1Rs, dynorphin, and substance P transcripts were not detected by in situ hybridization in the older mutants, a finding that was not surprising given the molecular paradigm used to make the mutants. There was, however, a moderate increase in striatal D2R and enkephalin mRNA. Receptor autoradiographic studies confirmed the lack of D1-class binding in the mutant striatum and in contrast to young pups that showed a reduction in D2-like binding, older hyperkinetic pups had a substantial increase in striatal D2-class binding, providing a neurochemical substrate to explain the hyperkinetic phenotype. Furthermore, in contrast to the downregulated DAT levels seen in very young pups, autoradiographic quantitation showed an increase in mutant striatal DAT density. In addition to the changes described in the striatopallidal and nigrostriatal pathways, upregulated dynorphin and substance P mRNA expression was also seen in the cortex. Failure of detection of NEO transcripts in mutant pup brain (by in situ hybridization) ruled out the possibility of significant mosaicism (with respect to the failure of *Cre*-mediated recombination within the brain) as an explanation for the longer survival period in older pups (Wong et al., in preparation). The NEO gene is bracketed by LoxP sites, and failure of *Cre*-mediated in vivo recombination should have resulted in detectable NEO transcripts. The lack of D1R mRNA, D1-class binding, striatal substance P, and dynorphin expression adds further support to the argument against mosaicism. The *tox-176* model provides compelling evidence in favor of the segregation hypothesis of D1 and D2 DA receptors, and the extensive neurochemical and behavioral changes described confirm the capacity of the developing brain for significant adaptation following selective cell injury.

The phenomenon of transcellular effects seen in cells intimately associated with cells in which intracellular DTA/*tox-176* production occurs has been explored in a number of additional studies. Targeted toxin-mediated Schwann cell ablation (Cole et al., 1994) resulted in a decrease in axonal diameter and intraneuronal neurofilament phosphorylation in addition to upregulated neurofilament density. Dramatic and widespread death of motoneurons was shown to occur during embryogenesis following activation of DTA expression in skeletal muscles, validating earlier studies suggesting a dependence of motor neurons on target-derived survival molecules (Grieshammer et al., 1998). Targeted ablation of cerebellar Purkinje cells during development results in transneuronal effects in a nearby population of neuronal precursors and mature neurons in the cerebellar cortex (Smeyne et al., 1995). In this study, the cerebellar external granular layer overlying the degenerate zone of Purkinje cells was shown to be markedly reduced in cell density due to a reduction in progenitor cell proliferation and an increase in apoptosis of committed granule cells.

4. FUTURE DIRECTIONS

The power of the DTA/*tox-176*-mediated cell ablation approach is evident from the large number of studies utilizing the method to examine fundamental questions in biology. The proven usefulness of the method in post-mitotic cells such as striatal projection neurons and cerebellar Purkinje cells is a major advantage. Future studies involving spatial and temporal control of *Cre* delivery will dramatically refine the system and enhance the usefulness of parental lines, such as the D1R/*tox-176* knock-in line in which dormant toxigenes can be activated in anatomically restricted cell populations thereby allowing a correlation between the loss of selected D1R-rich regions of the striatum and behavioral sequelae. Specifically, selective D1R-positive cell death may be achieved following stereotaxic *Cre* transgene delivery by using viral vectors. In addition, the ultimate generation of transgenic mice in which *Cre* production can be regulated by tetracycline derivatives will provide an exciting future refinement that will certainly enhance the biological power of the transgenic approach in understanding cell-cell interactions in the developing and mature organism.

REFERENCES

Accili D, Fishbourne CS, Drago J, Steiner H, Lachowicz JE, Park BH, (1996): A targeted mutation of the D3 dopamine receptor gene is associated with hyperactivity in mice. PNAS USA 93:19451949.

Aguila HL, Hershberger RJ, Weissman IL (1995): Transgenic mice carrying the diphtheria toxin A chain gene under the control of the granzyme A promoter: expected depletion of cytotoxic cells and unexpected depletion of CD8 T cells. PNAS USA 92:10192–10196.

Albin RL, Young AB, Penney JB (1989): The functional anatomy of basal ganglia disorders. Trends Neurosci 12:366–375.

Al-Shawi R, Burke J, Jones CT, Simons JP, Bishop JO (1988): A Mup promoter-thymidine kinase reporter gene shows relaxed tissue-specific expression and confers male sterility upon transgenic mice. Mol Cell Biol 8:4821–4828.

Andrews TC, Weeks RA, Turjanski N, Brooks DJ (1998): Monitoring disease progression in presymptomatic and early Huntington's disease: D1and D2 ligand PET and the Unified Huntington's Disease Rating Scale. Abs Am Acad Neurol S50.002.

Barde Y-A (1989): Trophic factors and neuronal survival. Neuron 2:1525–1534.

Behringer RR, Mathews LS, Palmiter RD, Brinster RL (1988): Dwarf mice produced by genetic ablation of growth hormone-expressing cells. Gene Dev 2:453–461.

Borrelli E, Heyman RA, Hsi M, Evans RM (1988): Targeting of an inducible toxic phenotype in animal cells. PNAS USA 85:7572–7576.

Borrelli E, Heyman RA, Arias C, Sawchenko PE, Evans RM (1989): Transgenic mice with inducible dwarfism. Nature 339:538–541.

Braun RE, Lo D, Pinkert CA, Widera G, Flavell RA, Palmiter RD, et al. (1990): Infertility in male transgenic mice: disruption of sperm development by HSV-tk expression in postmeiotic germ cells. Biol Reprod 43:684–693.

Breitman ML, Clapoff S, Rossant J, Tsui LC, Glode LM, Maxwell IH, et al. (1987): Genetic ablation: targeted expression of a toxin gene causes microphthalmia in transgenic mice. Science 238:1563–1565.

Breitman ML, Rombola H, Maxwell IH, Klintworth GK, Bernstein A (1990): Genetic ablation in transgenic mice with an attenuated diphtheria toxin A gene. Mol Cell Biol 10:474–479.

Bruggeman LA, Xie HX, Brown KS, Yamada Y (1991): Developmental regulation for collagen II gene expression in transgenic mice. Teratology 44:203–208.

Burrows HL, Birkmeier TS, Seasholtz AF, Camper SA (1996): Targeted ablation of cells in the pituitary primordia of transgenic mice. Mol Endocr 10:1467–1477.

Caille I, Dumartin B, Le Moine C, Begueret J, Bloch B (1995): Ontogeny of the D1 dopamine receptor in the rat striatonigral system: an immunohistochemical study. Eur J Neuro 7:714–722.

Capecchi MR (1989): Altering the genome by homologous recombination. Science 244:1288–1292.

Clark AJ, Iwobi M, Cui W, Crompton M, Harold G, Hobbs S, et al. (1997): Selective cell ablation in transgenic mice expression E. coli nitroreductase. Gene Ther 4:101–110.

Clifford JJ, Tighe O, Croke DT, Sibley DR, Drago J, Waddington JL (1998): Topographical evaluation of the phenotype of spontaneous behaviour in mice with targeted gene deletion of the D1A dopamine receptor: paradoxical elevation of grooming syntax. Neuropharmacology 37:1595–1602.

Cole JS, Messing A, Trojanowski JQ, Lee VM (1994): Modulation of axon diameter and neurofilaments by hypomyelinating Schwann cells in transgenic mice. J Neurosc 14:6956–6966.

Davies SW, Turmaine M, Cozens BA, DiFiglia M, Sharp AH, Ross CA, et al. (1997): Formation of neuronal intranuclear inclusions underlies the neurological dysfunction in mice transgenic for the HD mutation. Cell 90:537–548.

Drago J, Gerfern CR, Lachowicz JE, Steiner H, Hollon TR, Love PE (1994): Altered striatal function in a mutant mouse lacking D1A dopamine receptors. PNAS USA 91:12564–12568.

Drago J, Padungchaichot P, Accili D, Fuchs S (1998a): Dopamine receptors and dopamine transporter in brain function and addictive behaviors: insights from targeted mouse mutants. Dev Neurosc 20:188–203.

Drago J, Padungchaichot P, Wong JYF, Lawrence AJ, McManus JF, Sumarsono SH, et al. (1998b): Targeted expression of a toxin gene to D1 dopamine receptor neurons by *Cre*-mediated site-specific recombination. J Neurosc 18:9845–9857.

Dzierzak E, Daly B, Fraser P, Larsson L, Muller A (1993): Thy-1 tk transgenic mice with a conditional lymphocyte deficiency. Int Immunol 5:975–984.

El-Ghundi M, George SR, Drago J, Fletcher PJ, Fan T, Nguyen T, et al. (1998) Disruption of dopamine D1 receptor gene expression attenuates alcohol-seeking behavior. Eur J Pharma 353:149–158.

Ferrante RJ, Kowall NW, Beal MF, Richardson EPJ, Bird ED, Martin JB (1985): Selective sparing of a class of striatal neurons in Huntington's disease. Science 230:561–563.

Forno LS (1982): Pathology of Parkinson's disease. In Movement Disorders; Marsden CD, Fahn S, Eds.; Butterworths: London; pp 25–30.

Garabedian EM, Roberts LJ, McNevin MS, Gordon JI (1997): Examining the role of Paneth cells in the small intestine by lineage ablation in transgenic mice. J Biol Chem 272:23729–23740.

Gerfen CR (1992): The neostriatal mosaic: multiple levels of compartmental organization. Trends Neurosci 15:133–139.

Gerfen CR, Engber TM, Mahan LC, Susel Z, Chase TN, Monsma FJ Jr., et al. (1990): D_1 and D_2 dopamine receptor-regulated gene expression of striatonigral and striatopallidal neurons. Science 250:1429–1432.

Ginovart N, Lundin A, Farde L, Halldin C, Backman L, Swahn CG, et al. (1997): PET study of the pre-and post-synaptic dopaminergic markers for the neurodegenerative process in Huntington's disease. Brain 120:503–514.

Gozes I, Glowa J, Brenneman DE, McCune SK, Lee E, Westphal H (1993): Learning and sexual deficiencies in transgenic mice carrying a chimeric vasoactive intestinal peptide gene. J Mol Neuro 4:185–193.

Graybiel AM (1990): Neurotransmitters and neuromodulators in the basal ganglia. Trends Neurosci 13:244–254.

Grieshammer U, Lewandoski M, Prevette D, Oppenheim RW, Martin GR (1998): Muscle-specific cell ablation conditional upon *Cre*-mediated DNA recombination in transgenic mice leads to massive spinal and cranial motoneuron loss. Dev Bio 197:234–247.

Guennoun R, Bloch B (1992): Ontogeny of D1 and DARPP-32 gene expression in the rat striatum: an in situ hybridization study. Mol Brain R 12:131–139.

Herrera PL, Huarte J, Zufferey R, Nichols A, Mermillod B, Philippe J, et al. (1994): Ablation of islet endocrine cells by targeted expression of hormone-promoter-driven toxigenes. PNAS USA 91:12999–13003.

Hornykiewicz O (1966): Dopamine (3-hydroxytyramine) and brain function. Pharm Rev 18:924–964.

Hughes AJ, Daniel SE, Kilford L, Lees AJ (1992): The accuracy of the clinical diagnosis of idiopathic Parkinson's disease; a clinic-pathological study of 100 cases. J Neurol Neurosci Psy 55:181–184.

Jaber M, Robinson SW, Missale C, Caron MG (1996): Dopamine receptors and brain function. Neuropharmacology 35:1503–1519.

Kane JM, Woerner MG, Lieberman JA (1988): Tardive dyskinesia: prevalence, incidence, and risk factors. J Clin Psychopharm 8:52–56.

Kaur S, Key B, Stock J, McNeish JD, Akeson R, Potter SS (1989): Targeted ablation of alpha-crystallin-synthesizing cells produces lens-deficient eyes in transgenic mice. Development 105:613–619.

Kendall SK, Saunders TL, Jin L, Lloyd RV, Glode LM, Nett TM, et al. (1991): Targeted ablation of pituitary gonadotropes in transgenic mice. Mol Endocr 5:2025–2036.

Korfhagen TR, Glasser SW, Wert SE, Bruno MD, Daugherty CC, McNeish JD, et al. (1990): Cis-acting sequences from a human surfactant protein gene confer pulmonary-specific gene expression in transgenic mice. PNAS USA 87:6122–6126.

Lakso M, Pichel JG, Gorman J, Sauer B, Okamoto Y, Lee E, et al. (1996): Efficient in vivo manipulation of mouse genomic sequences at the zygote stage. PNAS USA 93:5860–5865.

Lakso M, Sauer B, Mosinger BJ, Lee EJ, Manning RW, Yu S-H, et al. (1992): Targeted oncogene activation by site-specific recombination in transgenic mice. PNAS USA 89:6232–6236.

Le Mouellic H, Lallemand Y, Brulet P (1992): Homeosis in the mouse induced by a null mutation in the Hox-3.1 gene. Cell 69:251–264.

Lee P, Morley G, Huang Q, Fischer A, Seiler S, Horner JW, et al. (1998): Conditional lineage ablation to model human diseases. PNAS USA 95:11371–11376.

Lees AJ (1987): The Steele-Richardson-Olszewski syndrome (progressive supranuclear palsy). In Movement Disorders 2; Marsden CD, Fahn S, Eds.: Butterworths: London; pp 272–287.

Lem J, Applebury ML, Falk JD, Flannery JG, Simon MI (1991): Tissue-specific and developmental regulation of rod opsin chimeric genes in transgenic mice. Neuron 6:201–210.

Lewandoski M, Meyers EN, Martin GR (1997): Analysis of FGF8 gene function in vertebrate development. Cold S Harb 62:159–168.

Li Q, Karam SM, Gordon JI (1996): Diphtheria toxin-mediated ablation of parietal cells in the stomach of transgenic mice. J Biol Chem 271:3671–3676.

Lowell BB, S-Susulic V, Hamann A, Lawitts JA, Himms-Hagen J, Boyer BB, et al. (1993): Development of obesity in transgenic mice after genetic ablation of brown adipose tissue. Nature 366:740–742.

Maxwell F, Maxwell IH, Glode LM (1987): Cloning, sequence determination, and expression in transfected cells of the coding sequence for the tox-176 attenuated diptheria toxin A chain. Mol Cell Biol 7:1576–1579.

Messing A, Behringer RR, Hammang JP, Palmiter RD, Brinster RL, Lemke G (1992): P0 promoter directs expression of reporter and toxin genes to Schwann cells of transgenic mice. Neuron 8:507–520.

Minasi LE, Kamogawa Y, Carding S, Bottomly K, Flavell RA (1993): The selective ablation of interleukin 2-producing cells isolated from transgenic mice. J Exp Med 177:1451–1459.

Nance MA (1998): Huntington disease: clinical, genetic, and social aspects. J Geriat Ps 11:61–70.

Palmiter RD, Behringer RR, Quaife CJ, Maxwell F, Maxwell IH, Brinster RL (1987): Cell lineage ablation in transgenic mice by cell-specific expression of a toxin gene. Cell 50:435–443.

Pappenheimer AM Jr (1977): Diphtheria toxin. Ann R Bioch 46:69–94.

Peterson BS (1996): Considerations of natural history and pathophysiology in the psychopharmacology of Tourette's syndrome. J Clin Psy 57:24–34.

Quinn N (1994): Multiple system atrophy. In Movement Disorders 3; Marsden CD, Fahn S, Eds.; Butterworth-Heinermann, Ltd.: Oxford; pp 262–281.

Rajput AH, Rozdilsky B, Rajput A (1991): Accuracy of clinical diagnosis in parkinsonism: a prospective study. Can J Neur 18:275–278.

Raymond SM, Jackson IJ (1995): The retinal pigmented epithelium is required for development and maintenance of the mouse neural retina. Curr Biol 5:1286–1295.

Ross SR, Graves RA, Spiegelman BM (1993): Targeted expression of a toxin gene to adipose tissue: transgenic mice resistant to obesity. Gene Dev 7:1318–1324.

Rothwell JC, Obeso JA (1987): The anatomical and physiological basis of torsion dystonia. In Movement Disorders 2; Marsden CD, Fahn S, Eds.;Butterworths: London; pp 313–331.

Salomon B, Lores P, Pioche C, Racz P, Jami J, Klatzmann D (1994): Conditional ablation of dendritic cells in transgenic mice. J Immunol 152:537–548.

Salomon B, Maury S, Loubiere L, Caruso M, Onclercq R, Klatzmann D (1995): A truncated herpes simplex virus thymidine kinase phosphorylates thymidine and nucleoside analogs and does not cause sterility in transgenic mice. Mol Cell Biol 15:5322–5328.

Sato M, Tada N (1995): Preferential expression of osteocalcin-related protein mRNA in gonadal tissues of male mice. Bioc Biop R 215:412–421.

Sauer B (1993) Manipulation of transgenes by site-specific recombination; use of *Cre* recombinase. In Methods in Enzymology: Guide to Techniques in Mouse Development Wassarman PM, DePamphilis ML, Eds.; Academic Press: San Diego; Volume 225:890–900.

Sauer B, Henderson N (1988a): The cyclization of linear DNA in *Escherichia coli* by site-specific recombination. Gene 70:331–341.

Sauer B, Henderson N (1988b): Site specific DNA recombination in mammalian cells by the *Cre* recombinase of bacteriophage P1. PNAS USA 85:5166–5170.

Sibley DR, Hollon TR, Grinberg A, Huang SP, Drago J, Westphal H (1998): Progress in the creation of a D5 dopamine receptor knockout mouse. Arch Pharm 358:supplement 2, SC 4.2.

Sibley DR, Monsma FJJ (1992): Molecular biology of dopamine receptors. Trends Phar 13:61–69.

Smeyne RJ, Chu T, Lewin A, Bian FS, S-Crisman S, Kunsch C, et al. (1995): Local control of granule cell generation by cerebellar Purkinje cells. Mol Cell Ne 6:230–251.

Smith DR, Striplin CD, Geller AM, Mailman RB, Drago J, Lawler CP, et al. (1998): Behavioural assessment of mice lacking D1A dopamine receptors. Neuroscience 86:135–146.

Steele JC, Richardson JC, Olszewski J (1964): Progressive supranuclear palsy. Arch Neurol 10: 333–359.

Stewart JD, Hussey D, Jones C, Houle S, Thomson J, Guttman M (1998): PET studies in presymptomatic and symptomatic Huntington's disease patients comparing [18F]fluorodeoxyglucose, [11C]raclopride and [11C]SCH-23390. Abs Am Acad Neurol S50.003.

Surmeier DJ, Eberwine J, Wilson CJ, Cao Y, Stefani A, Kitai ST (1992): Dopamine receptor subtypes colocalize in rat striatonigral neurons. PNAS US 89:10178–10182.

Surmeier DJ, Reiner A, Levine MS, Ariano MA (1993): Are neostriatal dopamine receptors co-localized? Trends Neurosci 16:299–305.

Surmeier DJ, Kitai ST (1994): Dopaminergic regulation of striatal efferent pathways. Cur Op Neur 4:915–919.

Surmeier J, Song W-J, Yan Z (1996): Coordinated expression of dopamine receptors in neostriatal medium spiny neurons. J Neurosc 16:6579–6591.

Wallace H, Clarke AR, Harrison DJ, Hooper ML, Bishop JO (1996): Ganciclovir-induced ablation non-proliferating thyrocytes expressing herpesvirus thymidine kinase occurs by p53-independent apoptosis. Oncogene 13:55–61.

Wallace H, McLaren K, al-Shawi R, Bishop JO (1994): Consequences of thyroid hormone deficiency induced by the specific ablation of thyroid follicle cells in adult transgenic mice. J Endocr 143:107–120.

Waters CH, Faust PL, Powers J, Vinters H, Moskowitz C, Nygaard T, et al. (1993): Neuropathology of lubag (X-linked dystonia Parkinsonism). Movement D 8:387–390.

Watts RL, Mirra SS, Richardson EP (1994): Corticobasal ganglionic degeneration. In Movement Disorders 3; Marsden CD, Fahn S, Eds.; Butterworth-Heinermann Ltd.: Oxford; pp 282–299.

Xu M, Moratalla R, Gold LH, Hiroi N, Koob GF, Graybiel AM, et al. (1994): Dopamine D1 receptor mutant mice are deficient in striatal expression of dynorphin and in dopamine-mediated behavioral responses. Cell 79:729–742.

Yamaizumi M, Mekada E, Uchida T, Okada Y (1978): One molecule of diphtheria toxin fragment A introduced into a cell can kill the cell. Cell 15:245–250.

Zhou QY, Palmiter RD (1995): Dopamine-deficient mice are severely hypoactive, adipsic, and aphagic. Cell 83:1197–1209.

CHAPTER 11

ANTISENSE RNA-MEDIATED INHIBITION OF GENE EXPRESSION

HSIEN-YU WANG, CHRISTOPHER M. MOXHAM, and CRAIG C. MALBON*

1. INTRODUCTION

The use of antisense DNA and RNA reagents to suppress mRNA levels for targeted proteins provides degrees of freedom not available with many other strategies designed to eliminate specific genes or gene products (Miller et al., 1977; Haseloff, Gerlach, 1988). The utility of the antisense DNA/RNA strategy for study of cell signaling has been amply demonstrated over the past decade (Kleuss et al., 1991; Wang et al., 1992; Watkins et al., 1992; Kleuss et al., 1992; Moxham et al., 1993; Kleuss et al., 1993; Moxham et al., 1993; Ahnert-Hilger et al., 1994; Shih, Malbon, 1994; Kleuss et al., 1994; Watkins et al., 1994; Dean and McKay, 1995; Moxham and Malbon, 1996; Shih and Malbon, 1996; Bahouth et al., 1996). Molecular details of the precise mechanism(s) by which antisense RNA operates are not known, but empirically the expression of an antisense RNA with complementarity to a targeted mRNA results in a decline in the amount of expression of the translation product (i.e., target protein), as shown in Figure 11.1. The loss of this target protein may well be the basis for a disease or, in some cases outlined below, may prove therapeutic.

For antisense oligodeoxynucleotides (ODN), preparation of reagents requires little more skill than knowing the proper sequence for oligomer design and a commercial supplier for ODN synthesis and purification (Wagner, 1994). The use of antisense RNA approaches in vitro requires facility with simple techniques of molecular biology and can be readily accomplished through the constitutive expression of the properly designed antisense RNA molecule (see below) via a strong promoter. This simple approach to the non-regulatable expression of antisense RNA requires only that the constitutively active promoter

*Corresponding author.

Genetic Manipulation of Receptor Expression and Function,
Edited by Domenico Accili.
ISBN 0-471-35057-5   Copyright © 2000 Wiley-Liss, Inc.

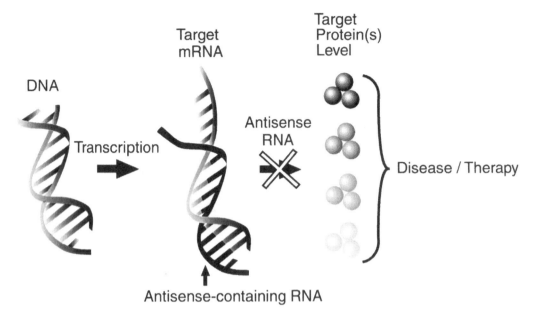

Figure 11.1. Antisense RNA suppresses the expression of a targeted protein. According to this paradigm, the transcription product of a specific, targeted gene (i.e., the mRNA of the targeted gene) hybridizes with a vector-expressed RNA molecule that contains an antisense region (antisense-containing RNA) to form a RNA:RNA duplex, which results in destruction of the duplex and/or inhibition of translation of the target mRNA. In either case, the expression of the antisense RNA leads to a decline and eventual elimination in the expression of the protein product of the targeted gene. The loss of a specific protein may create a model of a disease state or may well constitute a therapeutic approach to eliminating an aberrant protein or aberrant expression of a normal protein. See the text for details.

is "strong" enough to express sufficient levels of antisense RNA to hybridize stoichiometrically with the target RNA and that the promoter be functionally compatible with the cells to be targeted (Watkins et al., 1992; Watkins et al., 1994; Shih and Malbon, 1996; Bahouth et al., 1996). Gene promoters that can be "induced" afford an additional, exploitable capability, that is, expression of antisense RNA that can be turned "on" and again "off" in response to molecular signals. Exploitation of these features provides approaches to either induce or suppress antisense RNA. The inducible character of constructs enabling expression of antisense RNA is of particular utility in the suppression of mRNAs that encode proteins necessary for viability in cells or in the whole animal. Traditional knockout of genes by homologous recombination of such crucial targets provokes lethality in transgenic mice and consequently, no viable pups. Inducible antisense RNA transgenes are maintained "silently" in utero and then later can be turned "on" at birth or thereafter permitting study of the loss-of-function in viable transgenic pups. In the technical knockout (TKO) mice produced by inducible antisense RNA vectors, the additional time and expense of breeding to homozygosity required for traditional knockouts is avoided. The output of antisense RNA product from a single transgene under a strong promoter is sufficient to silence, via RNA:RNA hybridization, the mRNA for most of the protein targeted by the method to date.

Targeting multiple mRNAs with a single TKO RNA vector offers many advantages. Selecting a sequence antisense to the 5' untranslated region of a member of a homologous family of target gene products (at the nucleotide level) can create a vector that ablates the mRNA of solely one member by taking advantage of sequences unique to that family member. Selecting an antisense sequence to a region sharing homology/identity provides a vector that theoretically can ablate the mRNA of all of these family members. Thus one can target a single member of the heterotrimeric Gi family of alpha subunits targeting sequences in the 5'-untranslated region, whereas all three members of the Gi alpha family can be targeted through the selection of a sequence common to all, yet not common to alpha subunits of other families of heterotrimeric G-proteins. This extraordinary flexibility is as great as that of the *Cre*/LoxP system for multiple targets (Sauer, 1993), but with greatly reduced technical demand. Two approaches and their methodology will be detailed below. The strategies represent an evolution in antisense DNA/RNA technology—a technology with many skeptics in its earlier years but one that now enjoys widespread use in research applications and more recently in human therapeutics.

2. VECTOR-EXPRESSED ANTISENSE RNA FOR CELLS IN CULTURE

Vectors have been created that enable the expression of RNA antisense to mRNAs of targeted proteins. These antisense RNA sequences are by necessity expressed most often in the context of a larger RNA that acts as a "carrier". Small lengths of RNA suffer from instability upon expression, being degraded rapidly in comparison with oligodeoxynucleotides, modified or otherwise. Incorporating the antisense RNA sequence into a larger context of RNA enables accumulation that is not possible for expression of the short RNA sequences. A number of mammalian vectors have been used successfully to express antisense-RNA containing mRNA/RNA, including pCDNA3, pCMV5, pCW1, and pLNCX, to name but a few. For many of the applications in transgenic mice, antisense RNA expression vectors can be pre-screened in mammalian cells in culture. The screening permits verification of the nature of its construct and its ability to support expression of antisense RNA sequences embedded in other RNA molecules. In addition to testing the ability of the vector to express the antisense RNA (which can be accomplished via reverse-transcription, followed by PCR amplification), the ability of the antisense RNA to suppress the levels of the targeted mRNA as well as its translational product can be explored prior to the investment of time and expense in transgenic mice.

Although details of the mechanism(s) through which antisense RNA suppresses the levels of a targeted mRNA are still lacking, the generally accepted view is that the formation of the RNA:RNA duplex of antisense and sense RNA results in destruction of the duplex and/or inhibition of translation (see Fig. 11.1). Analysis of the relationship between the levels of antisense RNA and targeted mRNA suggests that the expression of antisense RNA leads to a combined loss of the antisense RNA itself as well as the targeted mRNA, favoring the notion that such RNA:RNA duplexes are rapidly destroyed (Moxham et al., 1993). Similar studies in cells stably transfected with an antisense RNA-expression vector have revealed the same general observations (Moxham et al., 1993)

3. VECTOR-EXPRESSED ANTISENSE RNA IN TRANSGENIC MICE

Application of antisense-RNA in tandem with transgenic mice offers an unparalleled opportunity for study of loss-of-function mutants of signaling molecules. The approach is much less technically demanding than homologous gene interruption and benefits from a much more rapid production of mice for study. Katsuki et al. (1992) demonstrated the antisense proof-of-concept early, succeeding in the transfer of an antisense RNA-producing "mini-gene" against myelin basic proteins (MBP) to fertilized mice zygotes. The transgenic mice developed a shiverer phenotype, the hallmark of pathophysiological demyelination of the CNS. Expression of antisense MBP mRNA was readily detected in the transgenic mice. Endogenous levels of MBP mRNA, the gene product, and of CNS myelination were shown to decline in the mutant mice.

Antisense RNA can be expressed in transgenic mice in many fashions, depending upon experimental demands and the ability to craft an expression vector with suitable characteristics to support the experimental plan. The expression of antisense RNA can be either constitutive or inducible. The domain of expression can vary from full expression in all cells to expression limited to specific cells or tissues of a transgenic mice. The considerations relevant to addressing these issues and the methods necessary to support the experimental goals are outlined in the sections below.

4. METHODS

Selection of Proper Sequences for Antisense RNA/DNA Applications

Application of antisense RNA to target the mRNA of a specific gene product necessitates the judicious selection of the sequence to be targeted. Almost all early examples of antisense DNA/RNA applications uniformly selected target sequences that included the ATG initiator codon, based upon theoretical concerns. Empirical determinations over the past several years suggest that the sequence(s) to be targeted actually can be selected from virtually any region of a particular mRNA. Successful ablation of signaling elements such as G-protein-linked receptors and G-proteins by ODNs, for example, has been achieved through targeting sequences in 5'-UTR, 3'UTR, as well as the ORF.

The "proper" size of antisense DNA/RNA for optimal use remains an overarching consideration in this strategy. ODNs <6–9 nucleotides in length appear incapable of forming stable heteroduplexes with their targeted RNAs. Thus a 9-nt is the lower limit for this parameter for ODNs involved in DNA:RNA hybridization (Agrawal S, 1996). Controversy exists as to the upper limit of the length of a DNA/RNA molecule that is optimal for antisense applications. A few examples have been published in which antisense sequences representing the entire cDNA of a gene product have been used. Although not to be considered a typical antisense RNA reagent, these inverted cDNAs represent the extreme case and also raise significant issues regarding their "specificity" for targeted mRNAs. The specificity of an inverted cDNA of a large expressed protein is limited by the simple empirical observation that, within relatively long reaches of DNA, units of homology/identity with other mRNAs that are not the target for the antisense reagents will be found. One could envision that by including the highly conserved regions of RNA encoding the GTP-binding capability of a G-protein alpha subunit, a number of G-proteins and other GTPases might be targeted inadvertently. A phenotype of interest is virtually guaranteed by use of a very long antisense RNA, which may abound with regions that target other unspecified target mRNAs. This obvious lack of specificity guarantees, however, that the probability of the phenotype observed being related to the targeting of solely the parent sequence as rather low. Selection of the proper sequence for antisense targeting that is not common, but rather unique with respect to target mRNA can be performed best by using the database of the *GeneBank* as a source. Exhaustive searches of the *GeneBank* are necessary to uncover all likely, inadvertent targets. By sliding the antisense sequence selection to more 3' or to more 5' regions in which homology to unwanted known targets is minimized/eliminated, one can reduce the chances of non-specific target acquisition to acceptable limits. The search for homology of unwanted targets, however, will be confined to known sequences in the database. Possible targets not yet sequenced are sure to exist and this real caveat demands careful attention in the interpretation of any study relying solely upon antisense RNA/DNA as a tool.

The best approach for the design of ODNs is minimalist, that is, designing the shortest sequence with the greatest specificity is optimal. Analysis of the theoretical optimal size for ODNs confirms the empirically derived data of those in the field. ODNs ranging between 20 and 30 nucleotides in length are optimal and a good starting point for the researchers naive to the strategy. The expense associated with ODN use for antisense approaches to cells in culture can be significant. Minimizing the length of the ODNs therefore can minimize unnecessary expense for the reagent.

Constitutive Expression of Antisense RNA in Cells in Culture

The best approach to the expression of RNA antisense to targeted gene products takes advantage of mammalian expression vectors of demonstrated capability. These vectors can use a variety of promoters, including that for early SV40, cytomegalovirus promoters, and LTRs, to provide constitutive expression of antisense RNA. By judicious selection of the cell type and vectorused, high-level, constitutive expression of antisense RNA can be obtained. For cells resistant to transfection by standard protocols, a retroviral construct can be used, which—when properly packaged—can provide high-efficiency introduction into a wide variety of cells. A broad spectrum of mammalian vectors have been created to suit most needs, and these typically can be obtained from commercial suppliers.

Retroviral Expression Vectors. Many cell lines demonstrate low efficiency for stable transfection when the traditional approaches are used to introduce plasmid DNA into mammalian cells in culture. In an effort to efficiently transfect cell lines resistant to transfection with calcium-phosphate precipitates of vector plasmid DNA, the pLNCX retroviral vector has been adapted to the task (Watkins et al., 1992; Watkins et al., 1994; Shih, Malbon, 1996; Bahouth et al., 1996). The pLNCXAS vector was first used to infect mouse F9 teratocarcinoma cells with a construct harboring an antisense sequence to the G-protein Gialpha2. This vector has proven useful in retroviral infection of a many cell types. Improvements in transfection approaches and reagents over the interim period has reduced to a much smaller number (cell types for which retroviral infection is required), although one can anticipate encountering such lines as more cell lines are explored as screens with which to evaluate the function of antisense RNA expression vectors targeted ultimately for use in transgenic mice.

Cytomegalovirus Promoter-Driven Constructs. The preponderance of literature suggests that mammalian expression vectors in which antisense sequences are driven by the cytomegalovirus (CMV) promoter are of the greatest utility. Harbored in the pCW1, pCMV5, pCDNA3, and similar expression vectors, antisense RNA sequences embedded in the context of a carrier RNA prove very effective at yielding high levels of expression of antisense RNA. Under selection with the NEO^r resistance gene, stable transfectants can be created in which constitutive expression of antisense RNA is robust and the expression of the targeted protein is abolished. In some cases in which the expression of antisense RNA targeted to an mRNA whose translation product is required for normal cell viability, constitutive expression would be avoided and the ability to induce the expression of antisense RNA would be advantageous.

Inducible Expression of Antisense RNA

The ideal for ablation of the production of a protein is an expression vector that can be induced to express the antisense RNA when desired. Stable transfectants can be isolated and propagated harboring an inducible construct targeting virtually any gene product. Constitutive expresses of antisense RNA throughout the selection process with a selectable marker, such as NEOr, may be accompanied by loss of cell proliferative capacity when the targeted protein is required for normal cell growth or may be accompanied by adaptive changes in the cells acting to rectify the loss of the targeted protein. To avoid loss of cell proliferative capacity and to minimize adaptive changes in the signaling pathways, several inducible promoters have been engineered into expression vectors, each basically offering the ability to be activated selectively and to induce the expression of antisense RNA, but with differing capabilities. Three examples of inducible promoter-driven vectors for antisense RNA are described that have been evaluated in our laboratory for the study of G-protein-linked receptor signaling and by other laboratories in other areas. The utility of the inducible strategy is described below. The list of inducible promoter constructs adapted for the production of antisense RNA includes heat shock, heavy metals, and hormone (steroid)-dependent activation of gene expression. The use of heat-shock promoters is rather embryonic for the purposes at hand. The heavy-metal promoters often are not strong enough to guarantee levels of antisense RNA expression that are high enough to surpass the endogenous levels of the target mRNA. The hormone (steroid)-dependent promoters are typically avoided when the study of cell-signaling pathways is the focus to preclude the unwanted halo effects often stimulated by the hormone treatment itself. Promoters that are inducible often suffer from being leaky, which compromises the advantage of the inducible strategy if expression of antisense RNA can not be operationally suppressed in vivo. Finally, the ability to adapt a mammalian expression vector for use in transgenic mice can be limited by the ease with which the promoter can be activated in vivo.

lac Operon-Driven Constructs. The *lac* operon provides an inducible eukaryotic expression system amenable to adaptation in antisense RNA strategies. Similar to the advantages provided by the ecdysone-inducible promoter described below, the advantages of the recently engineered *lac* operon system include rapid, robust expression of a gene or antisense transgene and reliable suppression of the target gene product by antisense RNA. The inducible character circumvents the problem of adaptive changes in the signaling pathways under study. The LacSwitch® system available from Stratagene offers a facile approach readily adapted for the purposes of inducible expression of antisense RNA. The system makes use of two vectors, one into which the antisense RNA construct is engineered and the other expressing high levels of the *lac* repressor. The lac repressor enables silencing of the first expression vector in the absence of the inducer. The LacSwitch vector into which the target gene or antisense construct is inserted contains multiple restriction sites to facilitate directional insertion. The repressor activity is expressed by addition of the inducer ligand IPTG into the medium of stable transfectants harboring both plasmids. The IPTG is added directly to the medium at 1–5 mM, and induction is measurable within 4–8 hr. For use in driving antisense RNA constructs, the suppression of target-gene product will reflect the relative half-life of the preexisting protein, following the destruction of the target mRNA by the RNA duplex formation with the antisense RNA. In our experience, the suppression of G-protein subunit expression requires 24–48 hrs post-induction with IPTG. Although not creating and instantaneous, "step-like" disappearance of the target protein, the system does offer an additional dimension to the use of inducible antisense RNA constructs.

p-IND® Promoter-Driven Constructs. The ecdysone-inducible promoter has been studied as an alternative approach for inducible antisense RNA expression. The ecdysone-inducible promoter is derived from *Drosophila melanogaster*. The promoter takes advantage of the pIND expression vector (commercially available from Invitrogen), which harbors five E/GREs. A companion plasmid (pVgRXR) is required, which expresses a modified nuclear receptor for the response element and completes the cellular complement for adaptation of this system to many purposes. Muristerone A is a very potent ecdysone agonist and can be used as an inducer in the stably transfected mammalian cells as well as recently in transgenic mice (see below). It must be determined first if muristerone A itself stimulates or alters the pathways of target cells (or mice) and that the promoter construct is not leaky in the absence of the inducer. Ecdysone administration to control mice appears to offer few effects (No et al., 1996). In cells in culture, the ecdysone agonist induces the expression of the antisense RNA to very high levels within 24 hr (not shown). Early results from the application of the ecdysone-sensitive promoter for suppression of G-protein subunits Gβ1 and Gβ2 by antisense RNA, suggest that expression of antisense RNA induced by the steroid provokes an inducible decline in the expression of the targeted subunit protein within 48-hr post-induction. RNA levels of Gαi2 remain unaffected and provide a useful point of reference for the specificity of the antisense RNA.

PEP-CK Promoter-Driven Constructs. As described in detail below, the goal of creating an inducible, tissue-specific transgene for use in transgenic animals focused interest upon the phosphoenolpyruvate carboxykinase (PEP-CK) promoter. In addition, a cell line in which to express and test the functional capabilities of the transgene prior to introduction into mice was required for screening. The utility of the pPEP-CK-AS vector to suppress expression of a signaling protein, in this case a heterotrimeric G-protein subunit, was evaluated first for study of $G_{i\alpha2}$ following transfection into FTO-2B rat hepatoma cells (Moxham et al., 1993; Moxham et al., 1993). The FT0-2B cells are liver-derived clones displaying cyclic AMP-inducible PEPCK gene expression and expressing both $G_{i\alpha2}$ and Gβ-subunits. The expression of RNA antisense to $G_{i\alpha2}$ in FTO-2B clones transfected with pPCK-ASG$_{i\alpha2}$ was detected by reverse transcription of total cellular RNA followed by PCR amplification. FTO-2B clones stably transfected with pPCK-ASG$_{i\alpha2}$ display wild-type levels of $G_{i\alpha2}$ expression in the absence of cyclic AMP, one partial inducer of the PEPCK gene expression. $G_{i\alpha2}$ levels declined >85% when these same cells were challenged with the cyclic AMP analog, 8-(4-chlorphenylthio)-cyclic AMP (CPT-cAMP) for 12 dy. For the FTO-2B clones stably transfected with the vector lacking the antisense sequence to $G_{i\alpha2}$, no change in $G_{i\alpha2}$ expression was observed. The expression of $G_{s\alpha}$ and $G_{i\alpha3}$ also was not changed in cells expressing the RNA antisense to $G_{i\alpha2}$, providing compelling proof that the antisense RNA sequence was specific for $G_{i\alpha2}$. The time elapsing between the induction of pPCK-ASG$_{i\alpha2}$ by CPT-cAMP and the decline of steady-state levels of $G_{i\alpha2}$ was approximately 6–12 days reflects the onset of the induction as well as the half-life of this subunit.

$G_{i\alpha2}$ is the most thoroughly studied member of the G_i family implicated in mediating the inhibitory adenylylcyclase pathway. Suppression of $G_{i\alpha2}$ expression in FTO-2B cells by introduction of the pPCK-ASG$_{i\alpha2}$ transgene largely attenuates receptor-mediated inhibition of adenylylcyclase (Moxham et al., 1993; Moxham, Malbon, 1996). Inhibition of forskolin-stimulated intracellular cyclic AMP levels by either somatostatin or the A1 purinergic agonist (-)-N[6]-(R-phenylisopropyl)-adenosine (R-PIA) is nearly abolished in transfectant cells in which expression of RNA antisense to $G_{i\alpha2}$ was first induced by CPT-cAMP for 12 days. FTO-2B clones that were stably transfected with the vector lacking the antisense sequence for $G_{i\alpha2}$ displayed a normal inhibitory adenylylcyclase response following a 12-day challenge with CPT-cAMP. These data demonstrated that $G_{i\alpha2}$ dominates as the mediator of the hormonal inhibition of liver adenylylcyclase. The pPEPCK-AS construct was shown to be a strong, inducible promoter capable of driving the production of an antisense RNA within the context of the PEPCK gene. The parent RNA molecule that contains the antisense sequence is remarkably stable, accumulates at high levels in response to induction for FTO-2B cells, and can be detected readily by RT-PCR by using the 5′ and 3′ flanking primers engineered into the antisense RNA cassette.

Transfection Protocol

1. Grow cells in one 100-mm cell-culture dish (log phrase: 1–2×10^6 cells)
2. To a 13-ml polypropylene tube, add the following (in order):

	Stock Solution	Per dish
H_2O		413.5 ml
Salmon-sperm DNA	5.0 mg/ml	4 ml
Neo$^+$ Plasmid	1.0 mg/ml	20 ml
$CaCl_2$	2 M	62.5 ml

3. To a tube containing 0.5 ml of $2\times$ HBS, add the above mixture drop-wise. At the same time, bubble in air for 30 s to mix.
4. Let the tube sit at room temperature for 30 min (the solution should become cloudy).
5. Gently add 1.0 ml of the above solution to the cell dish. Swirl gently to mix.
6. Incubate for 4 hr.
7. Aspirate off the medium, and wash the plate with 5.0 ml of DMEM. Then add 7.0 ml of fresh medium.
8. The following day, replace the old medium with new medium containing 400 mg/ml of G418 (Genticin®).
9. Change the medium every other day.

   Formulation of the $2\times$ HBS:

	Stock Solution	Volume
NaCl	2 M	6.75 ml
KCl	0.5 M	1.0 ml
Na_2HPO_4	100 mM	0.7 ml
Glucose	0.5 M	1.2 ml
HEPES	0.5 M	4.0 ml

   H_2O to 100 ml

   Adjust to pH 7.08, with NaOH; Filter and store at 4° C.

Notes. Controls for use of vector-expressed antisense RNA in vitro in stably transfected mammalian cells are of critical importance. Data gathered from studies with antisense ODNs are a useful guide for the selection of the target sequence used in vector-driven antisense RNA. The benefits of using vectors capable of expression of antisense RNA over antisense ODNs is obvious, substantial quantities of cells can be prepared in which antisense RNA has suppressed a targeted gene product. Biochemical analysis of the stably transfected clones is facile, avoiding miniaturization of assays for protein expression and the assays of signaling pathways confined to those with single-cell capability.

Prominent among pitfalls in this application is the variance among different clonal derivatives of a single transfection and the drift that sometimes occurs in clones expressing antisense RNA. At least 10 stable transfectant clones should be selected for expansion and characterization of both expression of the antisense RNA (utilizing unique primers and RT-PCR) and of the targeted protein (utilizing immunoblotting). The variance amongst clones from a single transfection can range from little suppression to virtual suppression of the targeted protein. The variance is encountered routinely, and multiple clones must be selected and characterized. Clones with optimal suppression are selected and used immediately for analysis of possible suppression of other related proteins, analysis of signaling pathways, and general characterization of the rate of growth.

Stable transfectants can display a "drift" in the amount of expression of antisense RNA with cell passage. The extent to which the clones can be frozen and thawed and can maintain the suppression phenotype is also quite variable. The drift in the cell population is rarely a general decline in the expression of the antisense RNA vector but rather the emergence of subclones that have lost the expression of the vector and overgrow the cells in which the antisense RNA continues to suppress expression of the targeted protein. Repeating the transfection and selection protocol entirely is more prudent than is an attempt to "rescue" cells from the drifting population. In our experience, antisense RNA expression and target protein suppression of many clones remains quite stable over time and cell passage. The extent to which suppression of a given protein is "toxic" to the cell and reduces cell growth or viability may dictate the extent to which the population drifts. The ability to suppress the expression of certain targets proteins displays a limit that may well indicate an important role in cell viability that cannot be sustained if suppressed further. Constitutive expression of RNA antisense to protein kinase A, protein kinase C, and GRKs, as well as protein phosphatases, has been used to probe the role of these proteins in the regulation of G-protein-linked receptors (Shih and Malbon, 1994, 1996).

Goals of Inducible, Tissue-Specific Ablation of Target Gene Products

Creating a strategy for suppressing the expression of a target protein in vivo must confront several formidable obstacles. The approach was designed for the broader use in targeting G-protein-linked receptors, other accessory signaling proteins, and general use, although the approach was first adapted to the elimination of a G-protein alpha subunit. The approach of using $G_{i\alpha 2}$-specific antisense RNA is far simpler than gene disruption by homologous recombination. To ensure accumulation of the antisense RNA in vivo, the target sequence was inserted in the 5' untranslated region of the rat phosphoenolpyruvate carboxykinase (PEPCK) mRNA. The PEPCK gene was selected for this work based on three considerations outlined earlier as prerequisites for a successful antisense RNA strategy: namely stability, inducibility, and tissue-specific expression (Moxham et al., 1993). With regard to stability, the 2.8-kb hybrid mRNA product of a transgene would be far more stable than a comparatively short-lived antisense RNA oligonucleotide fragment. With regard to inducibility, expression of PEPCK is controlled by several hormones, including glucagon (acting via cyclic AMP), glucocorticoids, and thyroid hormone, as well as insulin. Insertion of the antisense sequence within the PEPCK gene confers regulated expression of the desired antisense RNA sequences. Cyclic AMP coordinately increases the transcription rate of the gene as well as the stability of the mRNA, both features adding considerably to the utility of the transgene. Robust tissue-specific expression and "silence" of the transgene in utero were important considerations. The expression of the PEPCK gene is regulated developmentally, that is, initial appearance of the mRNA occurs at birth (Fig. 11.2). The mRNA of PEPCK, and thereby the transgene, is also highly abundant at birth, neonatal levels approach 0.5% to 1.0% of total cellular RNA. Targeted expression of the antisense RNA after birth obviates the problem of inducing a potentially lethal event by the suppression of a G-protein subunit in utero that might preclude viable transgenic pups. Finally, the tissue-specific expression is a critical feature of analysis of the many phenotypes. In the case of a traditional knockout, the loss of a specific gene product throughout an animal may yield a number of indirect effects that cloud analysis of the phenotype. Elimination of an important signaling element in the pituitary may yield pleiotropic effects that make study of the role of that protein in other sites almost impossible. The antisense RNA transgene can be engineered to include tissue-specific expression elements in the promoter. For the PEPCK-based transgene, we modified the promoter to restrict the domain of expression to essentially three target organs—liver, skeletal muscle, and adipose tissue (Fig. 11.2).

Other options for application of inducible gene expression for antisense RNA in transgenic mice are becoming available. As discussed above, the ecdysone-inducible gene expression system has proven useful in mammalian cells in culture. Ecdysone and its synthetic cogeners appear to have little effect on mammals, so the introduction of an ecdysone-responsive promoter into transgenic mice might offer advantages for inducible expression of antisense RNA. Although not yet applied to the antisense RNA strategy, expressing the ecdysone receptor can activate an integrated ecdysone-responsive promoter in transgenic mice administered with the hormone (No et al., 1996). In addition, significant experience had emerged with the use of tetracycline-responsive promoters in transgenic mice (Lee et al., 1998). The tet promoter has been used with success in transgenic mice (Saez et al., 1997). With a modified tetracycline-regulated system under the control of the neuron-specific enolase promoter, transgene expression has been targeted to specific regions of the mouse brain (Chen et al., 1998). Thus, many new applications for antisense RNA-induced suppression of signaling molecules will emerge as our understanding of tissue-specific and inducible promoters increases.

Screening of PEPCK Promoter-Driven Constructs In Vitro

As discussed above, the utility of the construct pPCK-AS to suppress expression of a signaling molecule can be readily evaluated after transfection into FTO-2B rat hepatoma cells (Moxham et al., 1993; Moxham, Malbon, 1996; Galvin-Parton et al., 1997; Guo et al., 1998). It can not be overstated that the screening of all pPCK-AS constructs in FTO-2B hepatoma cells is a necessary and worthwhile investment. The only constraints on the screening is, of course, that the product of

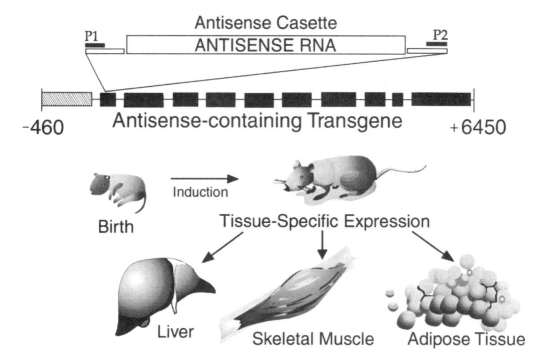

Figure 11.2. Inducible and tissue-specific expression of an antisense RNA transgene enables site-specific expression and avoidance of in utero lethality. Through engineering of the promoter region of the PEPCK gene, the pPCK-AS transgene was created for silent operation in utero (avoiding lethality for developmentally required gene products) and robust expression at birth in three target tissues_liver, skeletal muscle, and adipose. Analysis of the phenotypic changes of loss-of-function mutants in such a limited but readily studied subset of tissues facilitates detailed understanding of the loss-of-function accompanying the expression of the antisense RNA.

the targeted mRNA must be expressed in these liver-derived cells. The pPCK-AS/FTO-2B clone partnership provides only one extensively tested example for the use of inducible antisense RNA vectors ultimately targeted for use in transgenic mice. Other promoters with cell lines competent to test inducibility in vitro can be envisioned; for those cases in which sites of the mouse other than liver, skeletal muscle, and adipose tissues are to be targeted, new promoters and complementary cell lines for screening will be required. Examples of tissue-specific elements for lung, heart, and other tissues abound. Further testing of these promoter elements will the enable the creation of an array of antisense RNA vectors capable of inducible and tissue-specific expression throughout the mouse.

Creation of Transgenic Mice Harboring Inducible, Tissue-Specific Vectors

Transgenic lines of mice can be created at any number of commercial sites. Currently we make use of the University Transgenic Mouse Facility at

SUNY/Stony Brook, which uses standard techniques in transgenic work. The antisense vector pPCK-ASG$_{i\alpha2}$ (or an alternative) is excised free of vector sequences and purified prior to microinjection into the single-cell stage, pre-implantation embryos of either BDF1 or FVB strains of mice. The purity of the injected DNA is essential for high yield of transgenic mice. Short-cuts and attempts to use less-than-optimal DNA will impact heavily on the overall success of this part of the strategy. The embryos that have been successfully microinjected with the transgene then are transferred to pseudopregnant female mice. The offspring harboring the transgene are easily identified by PCR amplification. The transgenic nature of the mice identified by PCR amplification of the transgene-specific primers can be confirmed by subsequent Southern analysis using a pPCK-ASG$_{i\alpha2}$-specific probe. At the minimum, four separate founder lines should be created, identified by Southern analysis, and then used to breed an F1 generation of transgenic mice for use in experimentation. Necropsy and histological evaluation of transgenic mice and their littermates require availability of a competent mouse histologist. This service can be obtained for a fee from several commercial suppliers, including the Jackson Laboratory and Charles River Laboratory. The mice carrying the pPCK-ASG$_{i\alpha2}$ transgene, for example, display a runted phenotype that includes a marked reduction in total body mass (20%–40%), a sharp reduction in liver mass, and frank insulin resistance. More detailed analyses of the phenotype of transgenic mice harboring the inducible, tissue-specific promoter system for TKOs of G-protein alpha subunits have been reported earlier(Moxham et al., 1993a, 1993b, 1994, 1998; Moxham and Malbon, 1996; Galvin-Parton et al., 1997; Malbon et al., 1998).

For transgenic mice, the controls rely heavily on proof-of-concept developed by using cells in culture as a screen. It is not financially practical to create a series of transgenic mice in which missense and/or sense constructs are substituted for the antisense sequence. Our own experience of targeting several different G-protein subunits and exhaustively characterizing the derivative transgenic mice provides the proof-of-concept for the strategy.

A major limitation for studies of the transgenic mice is the cost incurred in the creation and maintenance of the mice colonies. The fee for creating transgenic mice depends upon the facility and whether the activity is subsidized. For academic settings the costs range between $2000 and $5000, with the commercial rates not very different from the costs of unsubsidized academic facilities. Four or five founder lines must be generated and maintained for each antisense vector construct. The animal husbandry for adequate production of mice for experimental purposes is 1500 to 2000 mice per construct. The mice that have been created using the pPEPCK transgene have been bred for >10 generations, are easily re-derived from the single cell embryos by microinjection, and are stable. The antisense RNA expression in vivo has surpassed our greatest expectation and remains restricted to the sites targeted in the initial design for the project now more than 8 years ago.

The pPEPCK-AS vector system is patented and licensed for use through the Research Foundation of SUNY/Stony Brook. For expression of antisense RNA in other target tissues, ample reports of analogous tissue-specific elements for the heart, regions of the CNS, and other tissues are easily retrieved from a search of the Medline.

Creation of Transgenic Mice Overexpressing Mutant Signaling Proteins

Parallel with the application of antisense RNA technology to achieve suppression of a signaling protein is the capacity to explore the other end of the spectrum. The strategy used to create loss-of-function mutants can be adapted to generate gain-of-function mutants in which a constitutively active form of a signaling protein (e.g., $G_{i\alpha2}$) may be expressed in a tissue-specific, inducible manner also. For example, we used this strategy to probe further the relationship between $G_{i\alpha2}$ and insulin action. $G_{i\alpha2}$-deficiency results in blunted glucose tolerance tests, loss of insulin-stimulated hexose transport in adipocytes, inability of insulin to recruit GLUT4 transporters to the plasma membrane, blunted glycogen synthase activation in both the and skeletal muscle of the transgenic mice, and enhanced expression of phosphotyrosine phosphatase 1b, but not *Syp* (Moxham and Malbon, 1996). This important linkage between insulin action and $G_{i\alpha2}$ was probed in the opposite direction, for investigating the effects, if any, of increased $G_{i\alpha2}$ activity on insulin action in transgenic mice. For heterotrimeric G-protein alpha subunits, it is possible to inactivate the endogenous GTPase responsible for the hydrolysis of bound GTP and its activation of the subunit itself. The cDNA encoding Q205L GTPase-deficient mutant $G_{i\alpha2}$ was engineered in a position 3' to and under the control of the PEPCK promoter as well as in a position upstream of the PEPCK 3' UTR, which enhances mRNA stability.

Embarking upon the creation of transgenic mice for antisense RNA vectors is a major decision. The limitations, pitfalls, and resources required deserve serious attention well in advance of the actual commitment (see Note No. 3). It must be highlighted, however, that advantages in speed and reduced cost associated with the use of inducible, antisense RNA transgenes as compared with homologous recombination to knock-out a target are quite significant.

ACKNOWLEDGMENTS

Studies in the authors' laboratory were supported by USPHS grant DK30111 from the NIDDK, National Institutes of Health. CMM is a recipient of National Research Service Award from T32 DK 07521, NIDDK, NIH.

REFERENCES

Agrawal S. Antisense Therapeutics; Humana Press: New Jersey; 1996.

Ahnert-Hilger G, Schafer T, Spicher K, Grund C, Schultz G, Wiedenmann, B (1994): Detection of G-protein heterotrimers on large dense core and small synaptic vesicles of neuroendocrine and neuronal cells. Eur J Cell 65:26–38.

Bahouth SW, Park EA, Beauchamp M, Cui X, Malbon CC (1996): Identification of a glucocorticoid repressor domain in the rat beta 1-adrenergic receptor gene. Rec Sig Trans 6:141–149.

Chen J, Kelz MB, Zeng G, Sakai N, Steffen C, Shockett PE, et al. (1998): Transgenic animals with inducible, targeted gene expression in brain. Molec Pharm 54:495–503.

Dean N, McKay R (1995): Inhibition of protein kinase C-alpha expression in mice after systemic administration of phosphorothioate antisense oligodeoxynucleotides. PNAS USA 91:11762–11766.

Galvin-Parton PA, Chen X, Moxham CM, Malbon CC (1997): Induction of Galphaq-specific antisense RNA in vivo causes increased body mass and hyperadiposity. J Biol Chem 272:4335–4341.

Guo JH, Wang HY, Malbon CC (1998): Conditional, tissue-specific expression of Q205L G-alpha12 in vivo mimics insulin activation of Jun N-terminal kinase and P38 kinase. J Biol Chem 273:16487–16493. Haseloff J, Gerlach WL (1988): Simple RNA enzymes with new and highly specific endoribonuclease activities. Nature 334:585–591.

Katsuki H, Kaneko S, Satoh M (1992): Involvement of postsynaptic G proteins in hippocampal long-term potentiation. Brain Res 581:108–114.

Kleuss C, Hescheler J, Ewel C, Rosenthal W, Schultz G, Wittig B (1991): Assignment of G-protein subtypes to specific receptors inducing inhibition of calcium currents. Nature 353:43–48.

Kleuss C, Scherubl H, Hescheler J, Schultz G, Wittig B (1992): Different beta-subunits determine G-protein interaction with transmembrane receptors [see comments]. Nature 358:424–426.

Kleuss C, Scherubl H, Hescheler J, Schultz G, Wittig B (1993): Selectivity in signal transduction determined by gamma subunits of heterotrimeric G proteins. Science 259:832–834.

Kleuss C, Schultz G, Wittig B (1994): Microinjection of antisense oligonucleotides to assess G-protein subunit function. Meth Enzym 237:345–355.

Lee P, Morley G, Huang Q, Fischer A, Seiler S, Horner JW, et al. (1998): Conditional lineage ablation to model human diseases. PNAS USA 95:11371–11376.

Malbon CC, Galvin-Parton PA, Wang HY, Moxham CM (1998: G-proteins regulating insulin action and obesity: analysis by conditional, targeted expression of antisense RNA in vivo. G-prot Dis 85–100Miller PS, Braiterman LT, Ts'o PO (1977): Effects of a trinucleotide ethyl phosphotriester, Gmp(Et)Gmp(Et)U, on mammalian cells in culture. Biochemistry 16:1988–1996.

Moxham CM, Hod Y, Malbon CC (1993): Gi alpha 2 mediates the inhibitory regulation of adenylylcyclase in vivo: analysis in transgenic mice with Gi alpha 2 suppressed by inducible antisense RNA. Dev Genet 14:266–273.

Moxham CM, Hod Y, Malbon CC (1993): Induction of G alpha i2-specific antisense RNA in vivo inhibits neonatal growth. Science 260:991–995.

Moxham CM, Malbon CC (1996): Insulin action impaired by deficiency of the G-protein subunit G ialpha2. Nature 379:840–844.

Moxham CM, Wang HY, Malbon CC (1994): G-proteins controlling differentiation, growth and development: analysis by antisense RNA/DNA Technology. Meth Neur 26:553–571. No D, Yao TP, Evans RM (1996): Ecdysone-inducible gene expression in mammalian cells and transgenic mice. PNAS USA 93:3346–3351.

Saez E, No D, West A, Evans RM (1997): Inducible gene expression in mammalian cells and transgenic mice. [Review] [54 refs]. Curr Opin B 8:608–616.

Sauer B (1993): Manipulation of transgenes by site-specific recombination: use of *Cre* recombinase. Meth Enzym 225:890–900.

Shih M, Malbon CC (1994): Oligodeoxynucleotides antisense to mRNA encoding protein kinase A, protein kinase C, and beta-adrenergic receptor kinase reveal distinctive cell-type-specific roles in agonist-induced desensitization. PNAS USA 91:12193–12197.

Shih M, Malbon CC (1996): Protein kinase C deficiency blocks recovery from agonist-induced desensitization. J Biol Chem 271:21478–21483.

Wagner RW (1994): Gene inhibition using antisense oligodeoxynucleotides. [Review] [53 refs]. Nature 372:333–335.

Wang HY, Watkins DC, Malbon CC (1992): Antisense oligodeoxynucleotides to GS protein alpha-subunit sequence accelerate differentiation of fibroblasts to adipocytes. Nature 358:334–337.

Watkins DC, Johnson GL, Malbon CC (1992): Regulation of the differentiation of teratocarcinoma cells into primitive endoderm by G alpha i2. Science 258:1373–1375.

Watkins DC, Moxham CM, Morris AJ, Malbon CC (1994): Suppression of Gi alpha 2 enhances phospholipase C signaling. Biochem J 299:593–596.

CHAPTER 12

DELIVERY OF ANTISENSE OLIGONUCLEOTIDES TO THE BRAIN TO ANALYZE RECEPTOR FUNCTION

SIMRANJIT KAUR and IAN CREESE

I. INTRODUCTION

The antisense strategy is a promising new tool for investigating the roles of numerous gene products in physiological systems and in pathological states. The recent advances in the field of genetics have made progress in antisense research rapid. The antisense technique is now being used to investigate the functions of numerous proteins such as neurotransmitter receptors, enzymes, G-proteins and cell-adhesion molecules among many other gene products and also in investigations into the localization of these proteins.

Antisense oligonucleotides have also been investigated as potential therapeutic agents, and indeed antisense technology was initially investigated in blocking viral replication and tumor cell growth (Hélène, and Toulmé, 1990). The licensing of fomivirsen by the Federal Drug Agency in 1998 certainly shows that these agents do have therapeutic potential. Fomivirsen, produced by the Isis Pharmaceutical Company, is beneficial in the treatment of cytomegalovirus (CMV) retinitis, a viral infection that can cause blindness in AIDS patients (Marwick, 1998). Isis also has a number of other antisense agents in various stages of clinical trials, including an intercellular cell adhesion molecule 1 (ICAM1) inhibitor, which is in Phase III clinical trials for Crohn's disease (Glover et al., 1997; Yacyshyn, 1998). Inhibitors of c-Raf kinase and Ha-Ras are also being investigated by Isis in the therapy of cancer (Xu et al., 1998). Besides the potential use of antisense agents in the therapy of various cancers (Bennett et al., 1996; Engelhard, 1998), other diseases may also be candidates—renal and cardiovascular diseases (Haller et al., 1998), viral infections such as HIV and influenza (Leiter et al., 1990; Lisziewicz et al., 1993),

Genetic Manipulation of Receptor Expression and Function,
Edited by Domenico Accili.
ISBN 0-471-35057-5   Copyright © 2000 Wiley-Liss, Inc.

and asthma. Nyce and Metzger (1997) found that antisense directed at adenosine A_1 receptors produced desensitization to challenges with adenosine or an allergen in rabbit models of asthma.

Molecular biologists are identifying new genes at a rapid pace and the function of the protein products of these genes is often unknown. The antisense technique provides a method with which to selectively decrease the levels of such novel proteins and to perhaps provide a clue to their functions. The premise behind the antisense strategy is simple and straightforward and, in theory, provides a selective tool with which to target functional proteins. Antisense oligonucleotides are synthesized molecules of single-stranded DNA and usually range from 15- to 30-bases long. The antisense oligonucleotide is designed to hybridize to a unique, complementary target mRNA sequence by Watson-Crick pairing. This hybridization prevents translation of the mRNA into its encoded protein. Antisense oligonucleotides, in theory, confer selectivity by virtue of their sequence-specificity and should be less toxic compared with pharmacological agents that are rarely specific for one site only and thus can produce adverse side-effects. Antisense sequences are also known to exist in the natural state and act as regulators of gene expression (Inouye, 1988; Kimelman, 1992).

The antisense technique was first used in 1978 when Zamecnik and Stephenson (1978) found that a synthetic oligodeoxynucleotide, added to a tissue culture of chick fibroblasts, inhibited the formation of new Rous sarcoma virus and prevented the transformation of the chick fibroblasts into sarcoma cells. Further developments occurred only after improvements were made in DNA sequencing (Maxam and Gilbert, 1977) and when automated solid-phase synthesis of the oligonucleotides was introduced (Caruthers, 1985; Letsinger and Lunsford, 1976). At present, in their unmodified form, oligonucleotides are easily synthesized in large amounts and of high purity by using automated methods (Cohen, 1992).

Antisense technology has been used extensively in in vitro studies and more recently in in vivo investigations. From initial studies in cell lines, these investigations have progressed to the central nervous system (CNS), where antisense agents have been used to reduce the expression of numerous receptors such as dopamine (Nissbrandt et al., 1995; Tepper et al., 1997; Weiss et al., 1996; Zhang and Creese, 1993), neuropeptide (Wahlestedt et al., 1993b), N-methyl-D-aspartate (Wahlestedt et al., 1993a), GABA (Morris et al., 1998), vasopressin (Landgraf et al., 1995), opioid receptors (King et al., 1997; Rossi et al., 1996; Sanchez-Blazquez et al., 1997) and even subunits of some of these receptors (Standaert et al., 1996; Zhao et al., 1996), of G-proteins (Plata-Salamàn et al., 1995; Shen et al., 1998), proteins such as SNAP-25 (Catsicas et al., 1996), neuropeptide Y (Akabayashi et al., 1994), enzymes such as tyrosine hydroxylase (Skutella et al., 1994) and tyrosine kinase (Bergan et al., 1994), transporters (Silvia et al., 1997) and early-onset genes such as *c-fos* (Chiasson et al., 1992, 1994; Heilig et al., 1993).

In addition to increased selectivity, antisense administration has another distinct advantage over pharmacological methods; it does not produce an up-regulation in receptor number following blockade of function, which is a problematic consequence of many pharmacological interventions (Weiss et al.,

1997). Also, for many recent studies, antisense agents have been used to target specific receptors for which selective and high-affinity antagonists are currently not available.

Antisense oligonucleotides also have an edge when compared with transgenic knockout animals, in which a gene is switched off in stem cells and results in the complete loss of protein expression in all cells of the maturing embryo. Antisense oligonucleotides can be administered directly into the brain region of interest and can be used to target pre- or post-synaptic receptors selectively (Tepper et al., 1997). Antisense knockdown is also reversible, with receptor levels returning to normal soon after the cessation of antisense treatment. Another advantage of the antisense technique over transgenic animals is that in the knockout animals, developmental changes may occur, which can compensate for the loss of the targeted receptor. These changes may have a far-reaching effect on steps downstream during maturation. Thus, any data obtained from transgenic animals is difficult to interpret.

In this chapter we will discuss the experimental parameters that should be taken into consideration when using antisense technology, and we will also briefly mention the advances made in elucidating some of the roles of dopamine D_2 receptors with antisense technology.

2. THE EXPERIMENTAL PARAMETERS IN ANTISENSE RESEARCH

For antisense oligonucleotides to be optimally effective in selectively hybridizing to their target mRNA sequences and to prevent the incidence of non-sequence specific toxic effects, numerous factors must be considered, These factors include the chemical structure of the antisense oligonucleotide molecule; efficient uptake into cells; stability at body temperature; the use of appropriate controls; the route of administration; and the optimal duration of administration, which is dependent on the turnover rate of the protein of interest.

Cellular Uptake and Mechanism of Action

Antisense oligonucleotides are thought to bind to an 80kD cell-surface DNA-binding site and are then internalized by cells by receptor-mediated endocytosis or non-selective pinocytosis (Loke et al., 1989; Yabukov et al., 1989). Yee et al. (1994) showed that after labeled oligonucleotides were intracerebroventricularly injected, they were detected largely in neurons but they were also found in macrophages, astrocytes, and pericytes.

Temsamani et al. (1994) showed that uptake into cells was linear for $\sim$8 h and then reached a maximum at 16 h when radiolabeled phosphorothioate oligonucleotides were used. The uptake was reported to be concentration-dependent; the oligonucleotides were taken up more efficiently at low concentrations, which is an indicator of a saturable transport system.

Phosphorothioate-modified oligonucleotides, which are more stable, have high affinity for cell-surface binding and uptake when compared with oligonucleotides with natural phosphodiester bonds (Zhao et al., 1993). When investigated in primary mouse spleen cells, fluorescein-labeled oligonucleotide backbones were found mainly in the cytoplasm with little staining found in the nucleus (Zhao et al., 1993), which corresponds to the finding made by Caceres and Kosik (1990). Temsamani et al. (1994) also reported greater accumulation of the radiolabeled phosphorothioate oligonucleotides in the cytoplasm as compared with the nucleus. In contrast, other reports have shown that when microinjected, phosphodiester or phosphorothioate oligonucleotides were localized mainly in the nucleus (Chin et al., 1990; Leonetti et al., 1991). The discrepancy between the two findings may be due to the injected oligonucleotides by-passing the intracellular compartments through which the internalized oligonucleotides pass (Zhao et al., 1993).

In an in vivo study, antisense oligonucleotides labeled with biotin or digoxigenin were injected into the lateral ventricle of rats, and the most intense staining was found in the areas in close proximity to the site of injection, as quickly as 15 min after intracerebroventricular injection (Yee et al., 1994). The labeled cells were reported to be neurons and both the cytoplasm and nucleus were stained, which corresponds to findings made by Sommer et al. (1993), who reported that the fluorescein-labeled oligonucleotides were found in both the cytoplasm and nuclei of striatal neurons. Szklarczyk and Kaczmarek (1995) also confirm this intracellular distribution. They infused radiolabeled oligonucleotides into the rat baso-lateral nucleus of the amygdala and detected the label in both the cytoplasm and nuclei.

The mechanism by which antisense oligonucleotides produce an effect is still unclear, but antisense oligonucleotides are thought to inhibit gene expression of the target protein by selectively binding to a target mRNA sequence and the resulting DNA:RNA heteroduplex prevents translation by stearic blocking (Liebhaber et al., 1984; Shakin and Liebhaber, 1986). The duplex is also a substrate for subsequent degradation by RNase-H (Walder and Walder, 1988). The cleavage is followed by further degradation and is thus, irreversible. Antisense oligonucleotides may also inhibit gene expression by interfering with ribosome binding and processing of mRNA, by interfering with mRNA conformation, or by mRNA splicing (for review see Cohen, 1992). Shen et al. (1998) intracerebroventricularly administered antisense directed at Gialpha2 in the mouse and found that although striatal Gialpha2 levels were reduced, there was no change in the levels of Gialpha2 mRNA, which indicates that the mechanism by which antisense reduces the levels of the target protein affects steps downstream of mRNA production. Phosphorothioate oligonucleotides can also form a triple-helix with double-stranded DNA (called the "antigene" strategy) and thus prevent transcription factors from binding and therefore, inhibiting transcription into RNA (Cooney et al., 1988; Moser and Dervan, 1987). Targets for this type of strategy are limited as they must be pure polypurine or polypyrimidine sequences (Gowers and Fox, 1999).

Phosphorothioates are toxic to some extent, so hybrid phosphodiester/phosphorothioate oligonucleotides have been investigated. These hybrids have a reduced number of phosphorothioate bonds but maintain a similar resistance to nucleases as the phosphorothioates. Ho et al. (1998) showed that the antisense effect produced by the most optimal oligonucleotide tested, a phosphodiester/phosphorothioate chimera, was RNase H-dependent, at least in the lateral septum. In tissue-culture studies, oligonucleotides targeted at sites downstream of the start codon, as were the oligonucleotides used by Ho et al. (1998), were found to act via RNase-H activation (Chiang et al., 1991; Ho et al., 1999).

The Design of Antisense Oligonucleotides

Antisense oligonucleotides consist of a series of bases that are arranged in a specific order to permit optimal hybridization to the mRNA sequence of the protein of interest. The antisense oligonucleotide sequences are cross-checked against GenBank to ensure that they do not bind at other sites and that they are selective only for their target sequence.

There are a number of chemical classes of oligonucleotides available, including the chemically unmodified phosphodiesters; the negatively charged phosphorothioates, in which one of the non-bridging oxygen atoms is replaced with a sulfur atom; the non-ionic methylphosphonates; conjugated oligodeoxynucleotides; the alpha-oligomers; and the end-capped oligodeoxynucleotides with phosphorothioate groups (Cohen, 1992). Recently a number of hybrids or chimeras have been developed, which include oligonucleotides with a mixed phosphodiester-phosphorothioate backbone (Ho et al., 1998).

Phosphodiester oligonucleotides are subsceptible to rapid degradation by exonucleases and endonucleases (Szklarcyzk and Kaczmarek, 1995; Wahlestedt, 1994; Whitesell et al., 1993) and only have a half-life in serum of ~20 min (Sands et al., 1994) and intracellular phosphodiester oligonucleotides are partially degraded within 4 h of cell culture (Zhao et al., 1993). However, the phosphodiesters are less likely to produce toxic side effects.

Phosphorothioate oligonucleotides are more stable due to improved nuclease resistance but they have reduced ability to activate RNase H-mediated cleavage of target mRNA (Gao et al., 1992; see review Ulmann and Peyman, 1990). In serum and in cell culture, phosphorothioates have a half-life of over 12 h (Mirabelli and Crooke, 1993; Sands et al., 1994), and they have also been recovered intact at least 24 h after administration into rat brains (Grzanna et al., 1998; Szklarcyzk and Kaczmarek, 1995; Whitesell et al., 1993). However, phosphorothioates have been shown to produce non-sequence specific toxic effects such as weight loss (Ho et al., 1998). Phosphorothioates bind to numerous proteins in a non-sequence specific manner, which may underlie their propensity to produce adverse effects (Brown et al., 1994; Guvakova et al., 1995).

Hybrids are now being designed that allow for greater stability and less toxicity. Ho et al. (1998) found that a chimera with phosphorothioate linkages added to a phosphodiester backbone was the most optimal at producing knockdown of corticotrophin releasing factor, CRF2 receptors and did not cause any overt illness or toxicity in the animals.

Other factors, which need to be considered when deciding which class of oligonucleotides to use, include the melting temperature of the DNA/RNA duplex or T_m, and this ideally should be high. Phosphorothioate and methylphosphonate modifications reduce the T_m for hybridization to target mRNA (Maher and Dolnick, 1988; Stein and Cohen, 1988; Stein et al., 1988). Also, the probability of secondary structures—such as a hairpin bend—being formed within the oligonucleotide molecule itself should be low, as these would impair the ability of the antisense molecule to linearize for action at the target sequence (Heilig, 1994).

In terms of cell uptake and tissue penetrability, the route of administration is also critical when considering the type of oligonucleotides to be utilized. Chemical modifications, as well as groups linked at the end of the oligonucleotides, will alter cellular uptake and intracellular distribution. Szklarczyk and Kaczmarek (1995) compared the pharmacokinetics of a phosphodiester and a phosphorothioate oligonucleotide by infusing them into the amygdala and reported that the phosphorothioates were more stable and had greater penetrability into cells, whereas the phosphodiesters showed greater tissue retention. Thierry and Dritschillo (1992) reported that in a leukemia cell line, end-capped phosphorothiates showed increased cellular uptake compared with the fully modified phosphorothioate analogs but were as stable.

Hydrophobic groups and polycations, such as polylysine, can also be attached to the oligonucleotide backbones to improve cellular uptake (Gewirtz et al., 1996; Hélène and Toulmè 1990). When encapsulated in a liposome, the cellular accumulation of end-capped phosphorothioates was found to be greater and localization was observed in the cytoplasm and nucleus (Thierry and Dritschillo, 1992). A similar finding was made by Alahari et al. (1998), who reported that when a liposomal transfection agent or a liposomal preparation was used, the antisense effect was potentiated.

Another factor that determines specificity is the length of the oligonucleotide sequence. It has been estimated that a sequence of about 12 bases could be sufficient to provide a unique sequence, given the number of genes in the human genome (Tyler et al., 1998). Gao et al. (1992) reported that minimizing the number of phosphorothioate linkages in the backbone could be beneficial as shorter sequences (15–20 mer in length) minimized the non-sequence specific inhibition of RNase-H, which occurred at high oligonucleotide concentrations, while the shorter sequences retained hybridization to their complementary RNA sequences at 37°C.

In our laboratory, we utilize 19-mer phosphorothioate-modified oligodeoxynucleotides and have been successful in reducing the expression of dopamine-receptor subtypes (Tepper et al., 1997; Zhang and Creese, 1993).

Route of Administration and Optimal Duration of Administration

The main strategies that have been used for administering antisense oligonucleotides into the brain are to infuse the oligonucleotides into the cerebral ventricle by using micro-osmotic pumps or by injecting them focally into the ventricles or into the brain area of interest, as oligonucleotides do not cross the blood brain barrier (Agrawal et al., 1991). These strategies allow for the antisense oligonucleotides to penetrate the brain tissue. Intraventricular infusion has been utilized by a number of investigators, including this group, and has been successful in reducing the number of receptors throughout the brain (Ho et al., 1998; Nissbrandt et al., 1995; Szklarczyk and Kaczmarek, 1995; Zhang and Creese, 1993; Zhang et al., 1994). The main advantage of intraventricular infusion is that the effect of decreases in receptor number can be investigated in a number of brain regions. However, intracerebral injections allow for lower concentrations of the antisense agents to be utilized, thus also reducing the incidence of any toxic effects. Also, a more precise administration of the oligonucleotides (Silvia et al., 1994) can allow for the distinction between pre- and post-synaptic receptors (Tepper et al., 1997) and help in elucidating the roles of the receptors in individual brain areas. We administer our antisense agents by intracerebroventricular infusion using mini-osmotic pumps and by intracerebral injections directly into the striatum or the substantia nigra with success (Tepper et al., 1997; Zhang and Creese, 1993).

The inhibitory effect of injected antisense oligonucleotides on receptor expression is short-lasting, due to the relatively short turnover rate of many receptors. Thus, to effectively decrease receptor number, the antisense oligonucleotides have to be administered frequently (Zhang et al., 1994; Zhou et al., 1994) or continuously (Weiss et al., 1996; Zhang and Creese, 1993), which is mainly because once the administration of antisense is ceased, the responses return to normal within a few days (Zhang and Creese, 1993; Zhang et al., 1994; Zhou et al., 1994). For example, the half-life of most G protein-linked receptors usually ranges from 2 to 5 days (Qin et al., 1994; Pasternak et al., 1980a, b), with the half-life of D_2 receptors ranging from 45 to 160 h, depending on the conditions in the study (Leff et al., 1984; Norman et al., 1987). Ho et al. (1998) showed that the duration of infusion of the oligonucleotides does have an effect, as when the authors increased the duration of infusion from 5 days to 9 days and the CRF2 receptor knockdown increased from 49% to 78%. Zhou et al. (1994) reported that although dopamine D_2 receptors were reduced after 1 day of treatment, the knockdown was almost complete 6 days after treatment began. In our laboratory, we have shown maximal catalepsy, a behavior associated with decreased dopamine D_2 receptor function, on the third day of D_2 antisense treatment (Zhang and Creese, 1993).

Davidkova et al. (1998) investigated a novel way of circumventing frequent or continuous administration by using circumventing an eukaryotic expression plasmid vector containing a cDNA sequence encoding an antisense RNA targeted to the dopamine D_2 receptor transcript. This allows for the antisense oligonucleotides to be produced intracellularly by normal processes. Single injections of the vector were made into the striata of mice and there was a long-lasting (>1 month) cataleptic response as well as a marked and long-lasting reduction in apomorphine-induced climbing behavior, which are responses indicative of a decrease in dopamine D_2 receptor function.

Controls for Antisense Studies

To determine if the oligonucleotides have produced a specific antisense effect, appropriate controls are essential and must be chosen with care. Controls that used commonly include mismatched oligonucleotides, in which a few bases are mismatched, and random oligonucleotides, which contain the same nucleotides as the antisense sequence but in a randomized order. Sense oligonucleotides are not recommended because they may hybridize to sequences for which they act as the antisense counterpart (Brysch and Schlingenspein, 1994; Wahlestedt, 1996). The nucleotide sequences of the controls should be cross-checked in GenBank to ensure that these sequences do not inadvertently have antisense properties at other known sequences. However, hybridization to sites that have yet to be sequenced cannot be totally discounted, at least not until the complete genome is sequenced. In the studies done in this laboratory, random oligodeoxynucleotides have been utilized and shown to be devoid of any apparent sequence-specific effects.

3. THE APPLICATION OF ANTISENSE TECHNOLOGY IN DOPAMINE-RECEPTOR RESEARCH

Molecular cloning techniques have identified five distinct dopamine receptors (Bunzow et al., 1988; Sokoloff et al., 1990; Sunahara et al., 1990, 1991; Van Tol et al., 1991). These dopamine receptors are subdivided into two classes based on similarities in amino acid sequences, G-protein associations, and affinity for pharmacological agents. The D1-like receptor family consists of D_1 and D_5 receptors, whereas the D2-like receptor family consists of D_2, D_3 and D_4 receptors (Civelli et al., 1993; Civelli, 1995).

It is difficult to conclusively investigate the functions of the receptors in the D2-like class due to a lack of potent and selective drugs that differentiate between them. There is a high degree of homology between the receptors in the D2-like receptor family, and most pharmacological agents that bind to the D_2 receptor also have affinity at the D_3 and D_4 receptors. In this laboratory we utilize antisense technology to distinguish between receptors in the D2-like receptor class.

Antisense Knockdown of Dopamine D_2 Receptors

Antisense Knockdown of Postsynaptic D_2 Receptors. The functional roles of the D_2 receptors have been investigated with antisense technology more so than any of the other dopamine receptors. D_2 receptors are abundant in the striatum, limbic regions, such as the nucleus accumbens and olfactory tubercle; they are thought to be involved in motivation, movement, and emotion. D_2 mRNA is also expressed by enkephalin-containing neurons and by cholinergic neurons (Bunzow et al., 1988; Civelli et al., 1993; Sibley and Monsma, 1992; Le Moine et al., 1990). There are no fully selective agents for the D_2 receptors at present, as drugs acting at the D_2 receptor also have affinity for the D_3 and D_4 receptors. A novel way to determine the functional roles of these receptors is by the use of antisense technology. Successful D_2 receptor knockdown has been achieved in a number of rodent brain studies carried out by this laboratory and others (Davidkova et al., 1998; Qin et al., 1995; Wahlestedt, 1996; Silvia et al., 1994; Tepper et al., 1997; Weiss et al., 1993; Zhang and Creese, 1993; Zhou et al., 1994).

In vivo administration of D_2 antisense inhibited only the behaviors mediated by D_2 receptors and not those mediated by the other dopamine receptors or by muscarinic receptors (Wahlestedt, 1996; Zhang and Creese, 1993, Zhou et al., 1994; Qin et al., 1995). D_2 antisense administered in vivo to unilaterally striatally 6-hydroxydopamine (6-OHDA)-lesioned mice reduced the response to the D_2 receptor agonist, quinpirole (Weiss et al., 1993; Zhou et al., 1994). Zhou et al. (1994) reported that the administration of D_2 antisense inhibited the rotations induced by the dopamine D_2 receptor agonist but had no effect on the rotations produced by a D_1 receptor agonist, SKF 38393, or a muscarinic receptor agonist, oxotremorine, in unilaterally 6-OHDA striatally lesioned mice. These observations indicate that the knockdown was specific to dopamine D_2 receptors.

Zhang and Creese (1993) infused a D_2 antisense 19-base phosphorothioate oligodeoxynucleotide (10 μg/μl), corresponding to codons 2–8 of the D_2 mRNA (5′-AGGACAGGTTCAGTGGATC-3′) (Bunzow et al., 1988), unilaterally into the lateral ventricle using osmotic mini-pumps over a three-day period. As the control, a 19-mer phosphorothioate-modified random oligodeoxynucleotide containing the same bases but in a randomized order (5′-AGAACGGCACT-TAGTGGGT-3′) was used. A 48% reduction in the number of D_2 receptors was demonstrated by saturation analysis after the antisense treatment, and the largest decreases occurred in the nucleus accumbens (72% reduction) compared with the striatum (~50% reduction). Catalepsy and locomotor activity measurements were taken to correlate the loss of receptor with functional deficits and a time-dependent increase in the cataleptic response, and a reduction in spontaneous motor activity was observed over the three-day administration period. Locomotor activity after quinpirole treatment was also decreased in the antisense-treated animals, whereas responses to SKF 38393 were not affected. These data show that the knockdown was specific to D_2 receptors (Zhang and Creese, 1993). In a more recent study, the involvement of D_2 receptors was investigated in the response to amphetamine in rats. It was observed that D_2 knockdown had no effect on the stereotypy response produced by high doses of amphetamine, while reducing the locomotor response to quinpirole. This finding was ascribed to the mechanism of action of amphetamine, that is, it increases dopamine release, which also acts at D_1 receptors in addition to D_2 receptors (Zhang et al., in press).

Qin et al. (1995) administered D_2 antisense oligodeoxynucleotides into the lateral ventricle as chronic injections, every 12 h for up to 8 days and found that there was a marked inhibition of D_2 receptor mediated behaviors but only a small reduction in the number of dopamine D_2 receptors in the mouse striatum. Administering the D_2 antisense after treatment with an irreversible D_2-receptor antagonist, fluphenazine-N-mustard (FNM) showed that the D_2 antisense reduced the rate of recovery of D_2 receptors in the mouse striatum as well as the restoration of normal motor activity after cessation of FNM treatment. Weiss and colleagues (Qin et al., 1995; Weiss et al., 1993; Zhou et al., 1994) find that after 6 days of D_2 antisense treatment in 6-OHDA lesioned mice, there is a minimal reduction in D_2 receptors but a near complete inhibition in quinpirole-induced locomotor activity. The authors conclude that there is a small pool of newly synthesized and functional receptors and antisense treatments preferentially target this pool of receptors (Qin et al., 1995).

In an in vitro study, a D_2 receptor antisense oligonucleotide was administered to a primary culture of rat pituitary cells and was reported to result in a reduced number of D_2 receptors and also to prevent the inhibition of adenylyl cyclase by the D_2 agonist, bromocriptine, as well as reducing prolactin mRNA levels (Valerio et al., 1994). Again, these effects were specific only to the D_2 antisense with the random oligodeoxynucleotide producing no reduction in receptor number.

The effects of site-specific D_2 antisense administration have also been investigated. A D_2 antisense oligonucleotide was either injected or infused unilaterally into the striatum of rats and was found to induce ipsilateral rotations after apomorphine or quinpirole treatment (Sun et al., 1996). These data strengthen the hypothesis that post-synaptic dopamine D_2 receptors are involved in motor activity. In a similar study, Rajakumar et al., (1997) infused D_2 antisense bilaterally into the striatum and observed a reduction in the stereotypic sniffing produced by high-dose apomorphine but had no effect on the behaviors seen after low-dose apomorphine administration. D_2 mRNA levels were found to be normal in the striatum and the substantia nigra, but a decrease in D_2 receptor binding was observed. The authors concluded that the response to high-dose apomorphine may occur through interactions at postsynaptic D_2 receptors, whereas the behaviors induced by low-dose apomorphine may be due to effects at the presynaptic D_2 receptors.

Antisense Knockdown of Presynaptic D_2 Receptors. Dopamine neurons in the substantia nigra express D_2 receptor mRNA and show D_2 receptor binding and electrophysiological effects (Brouwer et al., 1992; Mengod et al., 1992; Tepper et al., 1984). Dopamine agonists are reported to inhibit dopamine neuron firing (Akaoka et al., 1992; Bunney et al., 1978; Groves et al., 1975).

Tepper et al. (1997) investigated the role of pre-synaptic D_2 receptors by directly infusing D_2 antisense into the substantia nigra over a period of 3–6 days. A 50% reduction in nigral D_2 receptors was noticed on the treated side compared with the control side with receptor autoradiography, but there was no decrease in D_2 receptor binding in either striata. The D_2 antisense also decreased the apomorphine-induced inhibitory effect of dopamine on nigral dopamine cell firing. Tepper et al. (1997) also reported that D_2 autoreceptor knockdown increased somatodendritic and terminal excitability without affecting striatally induced inhibition of dopamine neurons, which involves GABA, thus showing that D_2 autoreceptors are expressed by nigrostriatal neurons at somatodendritic and axon terminal areas.

To investigate the role of D_2 autoreceptors in the behavioral response to cocaine, Silvia et al. (1994) administered D_2 antisense oligodeoxynucleotides unilaterally into the substantia nigra to target the autoreceptors and observed marked contralateral rotations after cocaine administration. The ability of sulpiride, a D_2 receptor antagonist, to increase electrically stimulated dopamine release was also markedly reduced, without any effect on basal striatal dopamine release, which the authors reasoned is consistent with a decrease in the D_2 autoreceptor number. There was an ~40% reduction in the number of nigral D_2 receptors on the treated side when compared with the untreated side. There were no alterations in the number of postsynaptic striatal D_2 receptors. The authors suggest that nigrostriatal D_2 autoreceptors play a direct role in the reduction of the motor response to cocaine, and the absence of spontaneous rotation in antisense-treated animals suggests that autoreceptor effects may be compensated for during normal behavior.

Histochemistry studies have shown that tyrosine hydroxylase (TH), the enzyme involved in dopamine synthesis, is increased in the ipsilateral substantia nigra after D_2 antisense treatment (Hadjiconstantinou et al., 1996; Kaur et al., in press; Sun et al., 1996), indicating increased dopaminergic activity after D_2 receptor knockdown.

4. THE LIMITATIONS OF ANTISENSE TECHNOLOGY AS IT STANDS CURRENTLY

Along with the obvious advantages of using antisense technology, such as specificity, there are a number of inherent disadvantages as well, especially as it stands currently. Poor blood-brain barrier penetrability and the risk of adverse effects, including decreased blood clotting and cardiovascular problems such as increased blood pressure and decreased heart rate (Gura, 1995), reduce the chances of antisense technology crossing over into the clinic successfully. However, the licensing of fomivirsen bodes well for the future of antisense oligonucleotides as therapeutic agents.

With the oligonucleotides available at present, direct administration into the CNS is required, but progress is being made and it is likely that systemically administered antisense molecules will be developed in the near future, which can cross the blood-brain barrier and produce effects centrally. The most potent and stable oligonucleotides used at present, the nuclease-resistant phosphorothioates, have been shown to produce non-sequence specific side-effects, but new chimeras are being investigated. The mixed phosphorothioate/phosphodiester backbone oligonucleotides appear promising as they produce a potent antisense effect and are as stable as the phosphorothioates.

The pharmacological phenomenon of "spare receptors" may also present a problem. The knockdown achieved by the antisense oligonucleotides usually ranges from 20% to 50% but from pharmacological studies often 70% to -90% receptor occupancy is needed to produce functional consequences (Hamblin and Creese, 1983; Meller et al., 1989; Saller et al., 1989). However, as reported by Qin et al., (1995), even a small reduction in the total population of receptors by antisense knockdown was sufficient to produce behavioral consequences.

Another problem could be due to the lack of selective high-affinity ligands for some of the sites investigated with antisense oligonucleotides. To quantify the receptor knockdown, receptor autoradiography or homogenate radioligand binding assays are often utilized, especially in the absence of any obvious behavioral deficits. However, a problem arises if there are no selective ligands, which bind to the receptor in question.

5. CONCLUSIONS

The antisense technique is an exciting new tool. With the current pace at which new genes are being identified in the various genome projects, antisense strategies are probably going to become more popular as tools in the investigations into the functions of these new genes.

REFERENCES

Agrawal S, Temsamani J, Tang JY (1991): Pharmacokinetics, biodistribution and stability of oligodeoxynucleotides phosphorothioates in mice. PNAS USA 88:7595–7599.

Akabayashi A, Wahlestedt C, Alexander JT, and Liebowitz SF (1994): Specific inhibition of neuropeptide Y synthesis in arcuate nucleus by antisense oligonucleotides suppresses feeding behavior and insulin secretion. Mol Brain R 21:55–61.

Akaoka H, Charlety P, Saunier CF, Buda M, Chouvnet G, (1992): Inhibition of nigral dopaminergic neurons by systemic and local apomorphine: possible contribution of dendritic autoreceptors. Neuroscience 49:879–892.

Alahari SK, DeLong R, Fisher MH, Dean NM, Viliet P, Juliano RL (1998): Novel chemically modified oligonucleotides provide potent inhibition of P-glycoprotein expression. J Pharm Exp 286, 419–428.

Bennett CF, Dean N, Ecker DJ, Monia BP. Pharmacology of antisense therapeutic agents, Cancer and Inflammation. In Methods in Molecular Medicine: Antisense Therapeutics; Agrawal S, Ed.; Humana Press: Totowa, NJ, 1996; pp 13–46.

Bergan R, Connell Y, Fahmy B, Kyle E, Neckers LM (1994): Aptameric inhibition of p210$^{bcr-abl}$ tyrosine kinase autophosphorylation by oligodeoxynucleotides of defined sequence and backbone structure. Nucl Acid R 22:2150–2154.

Brouwer N, Van Dijken H, Ruiters MH, Van Willingen JD, Ter Horst GJ (1992): Localization of dopamine D$_2$ receptor mRNA with non-radioactive in situ hybridization histochemistry. Neurosci L 142:223–227.

Brown DA, Kang SH, Gryaznov SM, DeDioniso L, Heidenreich O, Sullivan S, Xu X, et al. (1994) Effect of phosphorothioate modification of oligodeoxynucleotides on specific protein binding. J Biol Chem 269:26801–26805.

Brysch W, Schlingensiepen K-H (1994): Design and application of antisense oligonucleotides in cell culture, *in vivo* and as therapeutic agents. Cell Mol B 14:557–568.

Bunney BS, Aghajanian GK (1978): d-Amphetamine-induced depression of central dopamine neurons: evidence for mediation by both autoreceptors and a strio-nigral feedback pathway. Arch Pharm 304:255–261.

Bunzow JR, Van Tol HHM, Grandy DK, Albert P, Salon J, et al. (1988): Cloning and expression of a rat D$_2$ dopamine receptor cDNA. Nature 336:783–787.

Caceres A, Kosik KS (1990): Inhibition of neurite polarity by tau antisense oligonucleotides in primary cerebellar neurons. Nature 343:461–463.

Caruthers MH (1985): Gene synthesis machines: DNA chemistry and its uses. Science 230:281–285.

Catsicas M, Osen-Sand A, Staple JK, Jones KA, Ayala G, Knowles J, et al. Antisense blockade of expression; SNAP-25 in vitro and in vivo. In Methods in Molecular medicine: Antisense therapeutics Agrawal S, Ed. Humana Press: Totowa, NJ, 1996; pp 57–85.

Chiang MY, Chan H, Zounes MA, Freier SM, Lima WF, et al. (1991): Antisense oligonucleotides inhibit intercellular adhesion molecule I expression by two distinct mechanisms. J Biol Chem 266:18162–18171.

Chiasson BJ, Hooper ML, Murphy PR, Robertson HA (1992): Antisense oligonucleotide eliminates in vivo expression of c-fos in mammalian brain. Eur J Pharm 227:451–453.

Chiasson BJ, Armstrong JN, Hooper ML, Murphy PR, Robertson HA (1994): The application of antisense oligonucleotide technology to the brain: some pitfalls. Cell Mol N 14, 507–521.

Chin DJ, Green GA, Zon G, Szoka FC, and Straubinger RM (1990): Rapid nuclear accumulation of injected oligodeoxynucleotides. The New Biologist. 2, 1091–1100.

Civelli O (1995): Molecular biology of dopamine receptor subtypes. In Psychopharmacology: The Fourth Generation of Progress. (Eds. Bloom FE, and Kupfer DJ): Raven Press, New York. pp. 155–161.

Civelli O, Bunzow JR, and Grandy DK (1993): Molecular diversity of the dopamine receptors. Annu. Rev. Pharmacol. Toxicol. 32, 281–307.

Cohen J (1992): Oligonucleotide therapeutics. Trends Biotechnol. 10, 87–90.

Cooney M, Czernuszewicz G, Postel EH, Flint SJ, and Hogan ME (1988): Site-oligonucleotide bindng represses transcription of the human c-myc gene in vitro. Science, 241, 456–459.

Davidkova G, Zhou LW, Morabito M, Zhang SP, and Weiss B (1998): D2 dopamine antisense RNA expression vector, unlike haloperidol, produces long-term inhibition of D2 dopamine-mediated behaviors without causing up-regulation of D2 dopamine receptors. J Pharm Exp 285, 1187–1196.

Engelhard HH (1998): Antisense Oligodeoxynucleotide Technology: Potential use for the treatment of malignant brain tumors. Cancer Control: Journal of the Moffit Cancer Center 5, 163–170.

Gao W-Y, Han FS, Storm C, Egan W, and Cheng Y-C (1992): Phosphorothioate oligonucleotides are inhibitors of human DNA polymerases and RNase H: Implications for antisense technology. Mol. Pharmacol. 41, 223–229.

Gewirtz AM, Stein CA, and Glazer PM (1996): Facilitating oligonucleotide delivery: Helping antisense deliver on its promise. P NAS US 93:3161–3163.

Glover JM, Leeds JM, Mant TG, Amin D, Kisner DL, Zuckerman JE, Geary RS, Levin AA, Shanahan WR Jr (1997): Phase I safety and pharmacokinetic profile of an intercellular adhesion molecule-1 antisense oligodeoxynucleotide (ISIS 2302). J Pharm Exp 282, 1173–80.

Gowers DM, and Fox KR (1999): Towards mixed sequence recognition by triple helix formation. Nucl Acid R27, 1569–1577.

Groves PM, Wilson CJ, Young SJ, and Rebec GV (1975): Self-inhibition by dopaminergic neurons. Science, 190, 522–529.

Grzanna R, Dubin JR, Dent GW, Ji Z, Zhang W, Ho SP, Hartig PR (1998): Intrastriatal and intraventricular injections of oligodeoxynucleotides in the rat brain: tissue penetration, intracellular distribution and c-fos antisense effects. Brain Res. Mol Brain R 63, 35–52.

Gura T (1995): Antisense has growing pains. Science. 270, 575–577.

Guvakova MA, Yabukov LA, Vlodavsky I, Tonkinson JL, Stein CA, (1995): Phosphorothioate oligodeoxynucleotides bind to basic fibroblast growth factor, inhibit its binding to cell surface receptors and remove it from low affinity binding sites on extracellular matrix. J. Biol Chem. 270, 2620–2627.

Hadjiconstantinou M, Neff NH, Zhou LW, and Weiss B (1996): D2 dopamine receptor antisense increases the activity and mRNA of tyrosine hydroxylase and aromatic L-amino acid decarboxylase in mouse brain. Neurosci Lett. 217, 105–108.

Hamblin M, and Creese I (1983): Behavioral and radioligand binding evidence for irreversible dopamine receptor blockade by EEDQ. Life Sci. 32, 2247–2255.

Haller H, Maasch C, Dragun D, Wellner M, von Janta-Lipinski M, Luft FC (1998): Antisense oligodesoxynucleotide strategies in renal and cardiovascular disease. Kidney Int. 53, 550–558.

Hèlëne C, and Toulmè J-J (1990): Specific regulation of gene expression by antisense, sense and antigene nucleic acids. Biochem. Biophys. Acta. 1049, 99–125.

Heilig M, Engel JA, and Soderpalm B (1993): c-fos antisense in the nucleus accumbens blocks the locomotor stimulant action of cocaine. Eur J Pharm 236, 339–340.

Heilig M (1994): Antisense technology: Prospects for treatment of neuropsychiatric disorders. CNS Drugs. 1, 405–409.

Ho SP, Livanov V, Zhang W, Li J-H, and Lesher T (1998): Modification of phosphorothioate oligonucleotides yields potent analogs with minimal toxicity for antisense experiments in the CNS. Brain Res Mol Brain Res. 62, 1–11.

Ho SP, Bao Y, Lesher T, Conklin D, and Sharp D (1999): Regulation of the angiotensin type-1 receptor by antisense oligonucleotides occurs through an RNase H-type mechanism. Brain Res Mol Brain Res. 65, 23–33.

Inouye M (1988): Antisense RNA: its functions and applications in gene regulation-a review. Gene, 72, 25–34.

Kaur S, Creese I, and Tepper JM, Electrophysiological and Behavioural effects of dopamine receptor knockdown in the Brain. Antisense Technology in the CNS. ed. Leslie, Hunter and Robertson. Oxford University Press, Oxford, U.K. Submitted.

Kimelman D (1992): Regulation of eukaryotic gene expression by natural antisense transcripts. In Gene Regulation: Biology of Antisense RNA and DNA. Eds. Erikson, P. and Izant, J.G. Raven Press, New York. pp 1–10.

King M, Pan YX, Mei J, Chang A, Xu J, and Pasternak GW (1997): Enhanced kappa-opioid receptor-mediated analgesia by amtisense targeting of the sigma1 receptor. Eur J Pharm 331, R5–R6.

Landgraf R, Gerstberger R, Montkowski A, Probst JC, Wotjak CT, Holsboer F, and Engelman M (1995): V1 vasopressin receptor antisense oligodeoxynucleotide into septum reduces vasopressin binding, social discrimination abilities and anxiety-related behavior in rats. J. Neurosci. 15, 4250–4258.

Leff SE, Gariano R, and Creese I (1984): Dopamine receptor turnover rates in rat striatum are age-dependent. P NAS US. 81, 3910–3914.

Leiter JM, Agrawal S, Palese P, and Zamecnik PC (1990): Inhibition of influenza virus replication by phosphorothioate oligodeoxynucleotides. P NAS US. 87, 3430–3434.

Le Moine C, Normand E, Guitteny AF, Fouque B, Teoule R, and Bloch B (1990): Dopamine receptor gene expression by enkephalin neurons in the rat forebrain. P NAS US 87, 230–234.

Leonetti JP, Mechti N, Degols G, Gagnor C, and Lebleu B (1991): Intracellular distribution of microinjected antisense oligonucleotides. P NAS US 88, 2702–2706.

Letsinger RL, and Lunsford WB (1976): Synthesis of thymidine oligonucleotides by phosphate triester intermediates. J. Am. Chem. Soc. 98, 3655–3661.

Liebhaber SA, Cash FE, and Shakin SH (1984): Translationally associated helix-destabilizing activity in rabbit reticulocyte lysate. J. Biol. Chem. 259, 15597–15602

Lisziewicz J, Sun D, Metelev V, Zamecnik P, Gallo RC, and Agrawal S (1993): Long-term treatment of human immunodeficiency virus-infected cells with antisense oligonucleotide phosphorothioates. P NAS US. 90, 3860–3864.

Loke SL, Stein CA, Zhang XH, Mori K, Nakanishi M, Subasinghe C, Cohen JS, and Neckers LM (1989): Characterization of oligonucleotide transport into living cells. P NAS US 86, 3474–3478.

Maher LJ, and Dolnick BJ (1988): Comparative hybrid arrest by tandem antisense oligodeoxyribonucleotides or oligodeoxyribonucleoside methylphosphonates in a cell-free system. Nucl Acid R 16, 3343–3359.

Marwick C (1998): First "antisense" drug will treat CMV retinitis. JAMA. 280, 871.

Maxam AM, and Gilbert W (1977): A new method for sequencing DNA. P NAS US.74, 560–4.

Meller E, Bordi F, and Bohmaker K (1989): Behavioral recovery after irreversible inactivation of D_1 and D_2 dopamine receptors. Life Sci. 44, 1019–1026.

Mengod G, Villaro MT, Landwehrmeyer GB, Martinez-Mir MI, Niznik HB, Sunahara RK, Seeman P, O'Dowd BF, Probst A, and Palacios JM (1992): Visualization of dopamine D_1, D_2 and D_3 receptor mRNAs in human and rat brain. Neurochem. Int 20 (Suppl.): 33S–43S.

Mirabelli CK, and Crooke ST (1993): Antisense oligonucleotides in the context of modern molecular drug discovery and development. In Antisense Research and Applications. Eds. Crooke ST, and Lebleu B. CRC Press, Boca Raton. pp. 7–36.

Morris SJ, Beatty DM, and Chronwall BM (1998): $GABA_BR1a/R1b$-type receptor antisense deoxynucleotide treatment of melanotropes blocks chronic $GABA_B$ receptor inhibition of high voltage-activated Ca^{2+} channels. J. Neurochem. 71, 1329–1332.

Moser HE, and Dervan PB (1987): Sequence-specific cleavage of double helical DNA by triple helix formation. Science. 238, 645–650.

Nissbrandt H, Ekman A, Eriksson E, and Heilig M (1995): Dopamine D_3 receptor antisense synthesis in rat brain. NeuroReport. 6, 573–576.

Norman AB, Battaglia G, and Creese I (1987): Differential recovery rates of rat D_2 dopamine receptors as a function of aging and chronic reserpine treatment following irreversible modification: a key to receptor regulatory mechanisms. J. Neurosci. 7, 1484–1491.

Nyce JW, and Metzger WJ (1997): DNA antisense therapy for asthma in an animal model. Nature. 385, 721–725.

Pasternak GW, Childers SR, Snyder SH (1980a): Opiate analgesia: evidence for mediation by a subpopulation of opiate receptors. Science. 208, 514–516.

Pasternak GW, Childers SR, Snyder SH (1980b): Naloxazone, a long-acting opiate antagonist: effects on analgesia in intact animals and on opiate binding in vitro. J Pharm Exp 214, 455–462.

Plata-Salaman CR, Wilson CD, Sonti G, Borkoski JP, and French-Mullen JM (1995): Antisense oligodeoxynucleotides to G-protein alpha-subunit subclasses identify a transductional requirement for the modulation of normal feeding dependent on G alpha OA subunit. Brain Res. Mol Brain R 33, 72–78.

Qin ZH, Chen JF, Weiss B (1994): Lesions of mouse striatum induced by 6-hydroxydopamine differentially alter the density, rate of synthesis and level of gene expression of D1 and D2 dopamine receptors. J. Neurochem. 62, 411–420.

Qin Z-H, Zhou L-W, Zhang S-P, Wang Y, and Weiss B (1995): D_2 dopamine receptor antisense oligodeoxynucleotide inhibits the synthesis of a functional pool of D_2 dopamine receptors. Mol. Pharmacol. 48, 730–737.

Rajakumar N, Laurier L, Niznik HB, and Stoessl AJ (1997): Effects of intrastriatal infusion of D_2 receptor antisense oligonucleotide on apomorphine-induced behaviors in the rat. Synapse. 26, 199–208.

Rossi GC, Brown GP, Leventhal L, Yang K, and Pasternak GW (1996): Novel receptor mechanisms for heroin and morphine-6-beta-glucuronide analgesia. Neurosci Lett 216, 1–4.

Saller CF, Kreamer LD, Adamovage LA, and Salama AI. (1989): Dopamine receptor occupancy in vivo: measurement using N-ethoxycarbonyl-2ethoxy-1,2-dihydroquinoline (EEDQ). Life Sci. 45, 917–929.

Sanchez-Blazquez P, Garcia-Espana P, and Gaezon J (1997): Antisense oligodeoxynucleotides to opioid mu and delta receptors reduced morphine dependence in mice:
role of delta-2 opioid receptors. J Pharm Exp 280, 1423–1431.

Sands H, Gorey-Feret LJ, Cocuzza AJ, Hobbs FW, Chidester D, and Trainor GL (1994):
Biodistribution and metabolism of internally 3H-labeled oligonucleotides: I. Comparison of a phoshodiester and a phosphorothioate. Mol. Pharmacol. 45, 932–943.

Shakin SH, and Liebhaber SA (1986): Destabilization of messenger RNA/complemen-
tary DNA duplexes by the elongating 80 S ribosome. J Biol Chem 261, 16018-16025.

Shen J, Shah S, Hsu H, and Yoburn BC (1998): The effects of antisense to Gialpha2 on
opioid agonist potency and Gialpha2 protein and mRNA abundance in the mouse.
Brain Res Mol Brain Res. 59, 247–255.

Sibley DR, and Monsma FJ (1992): Molecular biology of dopamine receptors. Trends
Pharmacol. 131, 61–69.

Silvia CP, King GR, Lee TH, Xue Z-Y, Caron MG, and Ellinwood EH (1994): Intrani-
gral administration of D_2 dopamine receptor antisense oligodeoxynucleotides establishes a role for nigrostriatal D_2 autoreceptors in the motor actions of cocaine. Mol.
Pharmacol. 46, 51–57.

Silvia CP, Jaber M, King GR, Ellinwood EH, and Caron M (1997): Cocaine and amphetamine elicit differential effects in rats with a unilateral injection of dopamine
transporter antisense oligodeoxynucleotides. Neuroscience. 76, 737–747.

Skutella T, Probst JC, Jirikowski GF, Holsboer F, and Spanagel R (1994): Ventral
tegmental area (VTA): injection of tyrosine hydroxylase phosphorothioates antisense
oligonucleotide suppresses operant behavior in rats. Neurosci Lett 167, 55–58.

Sokoloff P, Giros B, Martres M-P, Bouthenet M-L, and Schwartz J-C. Molecular cloning
and characterization of a novel dopamine receptor (D_3): as a target for neuroleptics
(1990): Nature. 347, 146–151.

Sommer W, Bjelke B, Ganten D, and Fuxe K (1993): Antisense oligonucleotide to c-fos
induces ipsilateral rotational behaviour to d-amphetamine. NeuroReport. 5, 277–280.

Standaert DG, Testa CM, Rudolf GD, and Holingsworth ZR (1996): Inhibition of N-
methyl-D-aspartate glutamate receptor subunit expression by antisense oligodeoxynucleotides reveals their role in striatal motor regulation. J Pharm Exp 276, 342–352.

Stein CA, and Cohen JS (1988): Oligodeoxynucleotides as inhibitors of gene expression: A review. Cancer Res. 48, 2659–2688.

Stein OA, Subasinghe C, Shinozuka K, and Cohen JS (1988): Physicochemical properties of phosphorothioate oligodeoxynucleotides. Nucl Acid R 16, 3209–3221.

Sun B-C, Zhang M, Ouagazzal A-M, Martin LP, Tepper JM, and Creese I (1996):
Dopamine receptor function: An analysis utilizing antisense knockout in vivo, In
Pharmacological Regulation of Gene Expression in the CNS (Ed. Merchant, K.):,
CRC Press. pp. 51–78.

Sunahara RK, Niznik HB, Weiner DM, Stormann TM, Brann NR, Kennedy JL, Gelern-
ter JE, Rozmahel R, Yang Y, Israel Y, Seeman P, and O'Dowd BF (1990): Human
dopamine D1 receptor encoded by an intronless gene on chromosome 5. Nature.
347, 80–83.

Sunahara RK, Guan H-C, O'Dowd, BF, Seeman P, Laurier LG, George S, Torchia J, Van
Tol HHM, and Nizink HB (1991): Cloning of the gene for a human dopamine D_5 receptor with higher affinity for dopamine than D_1. Nature 350, 610–614.

Szklarczyk A, and Kaczmarek L (1995): Antisense oligodeoxyribonucleotides: stability
and distribution after intracerebral injection into rat brain. J. Neurosci. Methods. 60,
181–187.

Temsamani J, Kubert M, Tang J, Padmapriya A, and Agrawal S (1994): Cellular uptake of oligodeoxynucleotide phosphorothioates and their analogs. Antisense Res. Dev. 4, 35–42.

Tepper JM, Nakamura S, Young SJ, and Groves PM (1984): Autoreceptor-mediated changes in dopaminergic terminal excitability: effects of striatal drug infusions. Brain Res. 309, 317–333.

Tepper JM, Sun B-C, Martin LP, and Creese I (1997): Functional roles of dopamine D_2 and D_3 autoreceptors on nigrostriatal neurons analyzed by antisense knockdown in vivo. J. Neurosci. 17, 2519–2530.

Thierry AR, and Dritschilo A (1992): Intracellular availability of unmodified, phosphorothioated and liposomally encapsulated oligodeoxynucleotides for antisense activity. Nucl Acid R 20, 5691–5698.

Tyler BM, McCormick DJ, Hoshall CV, Douglas CL, Jansen K, Lacy BW, Cusack B, and Richelson E (1998): Specific gene blockade shows that peptide nucleic acids readily enter neuronal cells in vivo. FEBS Lett. 421, 280–4.

Uhlmann E, and Peyman A (1990): Antisense oligonucleotides: A new therapeutic principle. Chem. Rev. 90, 544–584.

Van Tol HHM, Bunzow JR, Guan H-C, Sunahara RK, Seeman P, Niznik HB, and Civelli O (1991): Cloning of a human dopamine D_4 receptor gene with high affinity for the antipsychotic clozapine. Nature 350, 614–619.

Valerio A, Belloni M, Gorno ML, Tinti C, Memo M, and Spano P (1994): Dopamine D_2, D_3 and D_4 receptor mRNA levels in rat brain and pituitary during aging. Neurobiol. Aging. 15, 713–719.

Wahlestedt C, Golanov E, Yamamoto S, Pich EM, Koob GF, Yee F, and Heilig M (1993a): Antisense oligodeoxynucleotides to NMDA-R1 receptor channel protect cortical neurons from excitotoxicity and reduce focal ischaemic infarctions. Nature. 363, 260–263.

Wahlestedt C, Pich EM, Koob GF, Yee F, and Heilig M (1993b): Modulation of anxiety and neuropeptide Y-Y1 receptors by antisense oligodeoxynucleotides. Science. 259, 528–531.

Wahlestedt C (1994): Antisense oligonucleotide strategies in neuropharmacology. Trends Pharmacol. Sci. 15, 42–46.

Wahlestedt C (1996): Antisense ëknockdowní strategies in neurotransmitter receptor research, In Antisense Strategies for the study of receptor mechanisms (Eds. Raffa, R. and Porreca, F.): Landes Company. pp. 1–10.

Walder RY, and Walder JA (1988): Role of RNase H in hybrid-arrested translation by antisense oligonucleotides. P NAS US 85, 5011–5015.

Weiss B, Zhou L-W, Zhang S-P, and Qin Z-H (1993): Antisense oligodeoxynucleotide inhibits D_2 dopamine receptor-mediated behavior and D_2 messenger RNA. Neuroscience. 55, 607–612.

Weiss B, Zhou L-W, and Zhang S-P (1996): Dopamine antisense oligodeoxynucleotides as potential novel tools for studying drug abuse, In Antisense Strategies for the study of receptor mechanisms (Eds. Raffa, R.B. and Poreca, F.). R.G. Landes, Austin, TX. pp. 71–91.

Weiss B, Zhang SP, and Zhou LW (1997): Antisense strategies in dopamine receptor pharmacology. Life Sci. 60, 433–55.

Whitesell L, Geselowitz D, Chavany C, Fahmy B, Walbridge S, Alger JR, and Neckers LM (1993): Stability, clearance and disposition of intraventricularly administered oligodeoxynucleotides: implications for therapeutic application within the central nervous system. P NAS US 90, 4665–4669.

Xu XS, Vanderziel C, Bennett CF, and Monia BP (1998): A role for c-Raf kinase and Ha-Ras in cytokine-mediated induction of cell adhesion molecules. J Biol Chem 273, 33230–33238.

Yabukov LA, Deeva EA, Zarytova VF, Ivanova EM, Ryte AS, Yurchenko LV, and Vlassov VV (1989): Mechanism of oligonucleotide uptake by cells: Involvement of specfic receptors? P NAS US 86, 6454–6458.

Yacyshyn BR (1998): Drug treatment for Crohn's disease. Lancet. 3527, 743.

Yee F, Ericson H, Reis DJ, and Wahlestedt C (1994): Cellular uptake of intracere-broventricularly administered biotin- or dixogenin-labeled antisense oligodeoxynu-cleotides in the rat. Cell Mol N 14:475–486.

Zamecnik PC, Stephenson ML (1978): Inhibition of Rous sarcoma virus replication and transformation by a specific oligodeoxynucleotide. P NAS US 75:280–284.

Zhang M, Creese I (1993): Antisense oligodeoxynucleotide reduces brain dopamine D_2 receptors: behavioral correlates. Neurosci Lett 161:223–226.

Zhang M, Ouagazzal A-M, Creese I (in press): Antisense knockout of brain dopamine D_2 and D_3 receptors and its effects on motor behaviors.

Zhang S-P, Zhou L-W, Weiss B (1994): Oligodeoxynucleotide antisense to the D_1 dopamine receptor mRNA inhibits D_1 dopamine receptor-mediated behaviors in normal mice and in mice lesioned with 6-hydroxydopamine. J Pharm Exp 271:1462–1470.

Zhao Q, Matson S, Herrera CJ, Fisher E, Yu H, Krieg AM (1993): Comparison of cellu-lar binding and uptake of antisense phosphodiester, phosphorothioate and mixed phosphorothioate and methylphosphonate oligonucleotides. Antisense R Dev 3:53–66.

Zhao T-J, Rosenberg HC, Chiu TH (1996): Treatment with an antisense oligodeoxynu-cleotide to the $GABA_A$ receptor $\gamma 2$ subunit increases convulsive threshold for β-CCM, a benzodiazepine "inverse agonists", in rats. Eur J Pharm 306:61–66.

Zhou L-W, Zhang S-P, Qin Z-H, Weiss B (1994): *In vivo* administration of an oligodeoxynucleotide antisense to the D_2 dopamine receptor messenger RNA in-hibits D_2 dopamine receptor-mediated behavior and the expression of D_2 dopamine receptors in mouse striatum. J Pharm Exp 268:1015–1023.

INDEX

Ablation
  protein
    antisense oligonucleotides for, 205–222
    antisense RNA for, 189–204
  toxin-mediated cell
    of cerebellar Purkinje cells, 183
    Cre-mediated system for, 178–183
    in dopamine receptors, 179–183
    herpes simplex virus-1 thymidine kinase
      for, 175–177
    nitroreductase for, 177
    prodrug paradigms for, 175–177
    of Schwann cells, 183
    targeted expression of diphtheria toxin for,
      175, 177–183
    targeted, in vivo, 175–188
    temporal and spatial specificity of, 175
    tissue-specific, 177–178
Adenosine receptors, type A_1, antisense against,
  205–206
Adenylyl cyclase pathway, inhibition of, by G
  proteins, 196–197
Age, of mice
  and behavioral phenotyping, 24
  classifications of, 24
Aggregation chimeras, 42
Alcohol
  abuse, economic cost of, 93
  mechanism of action
    caveats on studies of, 101–105
    and GABA type A receptor, 95t, 98–100,
      105
    gene targeting strategies in analysis of,
      93–110
    and kinases, 95t
    lipid-based theory of, 93
    mouse model of, 93–110
    mouse strains used in studies of,
      95t
    and neuropeptide Y, 95t, 97–98
    potential targets of, 94t

    protein-based theory of, 93–94
    and serotonin receptors, 95t, 96–97
Alkaline lysis kits, 13
Allelic series
  for *Fgfr1* analysis, 40–42, 41t
  generation of
    embryological manipulation for, 42
    ENU mutagenesis for, 42–43
    gene targeting for, 40
  for receptor phenotype analysis, 39–43
  in VP16-dependent binary system, 78–79
Alpha-oligomer oligonucleotides, 209
Alzheimer's disease, transgenic mouse models
  of, 24
Aminobutyric acid (GABA)
  antisense analysis of, 206
  inhibitory effects of, 98–99
  type A receptor
    alcohol and, 95t, 98–100, 105
    alpha 6 subunit of, 102
    anesthetics and, 95t, 100, 105
    in Angelman syndrome model, 102
    beta 3 subunit of, 95t, 100, 102
    gamma 2 subunit of, 95t, 99–100
    structure of, 98f, 98–99
Amphetamines, and dopamine receptors,
  213
Anesthetics, mechanism of action
  caveats on studies of, 101–105
  and GABA type A receptor, 95t, 100, 105
  gene targeting strategies in analysis of,
    93–110
  knockin approach to, 105–106
  lipid-based theory of, 94
  mouse model of, 93–110
  mouse strains used in studies of, 95t
  potential targets of, 94t
  protein-based theory of, 93–94
Angelman syndrome, GABA type A receptor
  in, 102
Antigene strategy, 208